Henry S. Truman

Household Hand-Book of Medicine

Henry S. Truman

Household Hand-Book of Medicine

ISBN/EAN: 9783337404840

Printed in Europe, USA, Canada, Australia, Japan

Cover: Foto ©berggeist007 / pixelio.de

More available books at **www.hansebooks.com**

HOUSEHOLD
HAND-BOOK

OF

MEDICINE.

BY

HENRY S. TRUMAN.

CHICAGO:
SPAULDING & CO.
1893.

CONTENTS.

                                                                                       Page.

INTRODUCTION 1
A FEW HINTS FOR CAREFUL CONSIDERATION.
    Importance of a Knowledge of Medicine..... 11
    "Household Remedies."................... 12
    The Family Medicine Chest................ 14
    Patent Medicines........................ 14
HYGIENE....................................... 16
HINTS ON NURSING.............................. 18
ACCIDENTS AND EMERGENCIES.................... 23
    Asphyxiation (Suffocation)................ 24
    Burns and Scalds....... 24
    Bruises 25
    Fractures or Broken Bones................ 26
    Cuts and Wounds........................ 27
    Choking................................. 29
    Concussion of the Brain.................. 30
    Dislocations. 30
    Drowning 32
    Eyes Accidents to the.................... 33
    Ear ache.. 33
    Fainting 34
    Freezing 34
    Griping................................. 35
    Gun Shot Wounds........................ 35

Prostration from Heat—Sunstroke..........35
Hydrophobia 36
Lock Jaw................................36
Bleeding from the Nose..................36
Poison Ivy..............................37
Sprains and Strains.....................37
Snake Bites.............................38
Stings of Insects.......................39
POISONS AND ANTIDOTES.......................40
COMMON DISEASES AND THEIR TREATMENT.........43
Asthma 45
Ague (Chills and Fever).................47
Biliousness 48
Boils...................................50
Bunions.................................51
Back ache...............................51
Bright's Disease........................52
Consumption 54
Cholera Morbus..........................57
Coughs and Colds (Bronchitis) 58
Chafing.................................59
Corns 59
Cholera.................................60
Chilblains..............................64
Chapped Skin............................65
Chicken Pox.............................65
Canker Sore Mouth.......................66
Constipation............................67
Catarrh.................................68
Croup 71
Diphtheria 74
Dyspepsia—Indigestion 76
Diarrhœa................................81
Dysentery...............................83

Inflammation of the Eyes	84
Erysipelas	86
Eczema	90
Felon	91
Fever sores	92
Gravel	92
Gout	94
Headache	95
Hives	97
Hay Fever	98
Hoarseness	101
Heart burn	102
Heart Disease	102
Hooping-Cough	104
Itch	106
Jaundice	108
Measles	110
Mortification—Gangrene	112
Neuralgia	113
Night Sweats	115
Piles	116
Influenza—La Grippe	117
Pimples	120
Pneumonia	122
Rheumatism	125
Ringworm	127
Scarlet Fever	127
Shingles	130
Sore Throat	131
Salt Rheum	132
Scrofula	133
Typhoid Fever	137
Toothache	140
Tumors	141

Ulcers.................................142
Warts.................................142
Worms......................... 143
MEMORANDA.

INTRODUCTION.

No branch of scientific research is of such vital moment to every human being as that of medicine. Health, how to preserve it, and how to bring about its return when stolen by disease, is the vital factor of success and happiness with us all.

In no other department of learning does there seem to be the degree of misapprehension or positive ignorance, among the people at large, as in this. This lamentable condition may have arisen from any of three causes and has probably been contributed to by all.

First:—Carelessness and lack of interest on the part of the general public.

Second:—The intense jealousy with which the medical profession, since the days of Esculapius, has guarded its knowledge and discoveries from the outside world; which has from time immemorial caused them to surround their lore with a bristling hedge of technical terms and phrases incomprehensible to any but the initiated.

Third:—The absence of reliable and intelligent popular treatises, adapted to the understanding of all and published at a price which would insure their thorough distribution.

It is the last of these reasons which has induced the author to prepare and launch this little work. In it he has endeavored to treat solely of the most common diseases and these as concisely and clearly as possible, omitting all scientific and technical phrases. Obscure and difficult ailments which can be recognized and handled only by a trained physician have been left out, as being only confusing elements; as well as the diseases peculiar to women which will be discussed in a separate volume soon to be published.

Lack of space has often compelled extreme brevity when the author would have preferred more extended mention, but he believes he has succeeded in including all of the more important points and the reader may rest assured that every statement has been fully verified by the latest and highest authorities.

With the hope that the little volume will in some slight measure realize his expectations, viz: to afford his readers help and some degree of understanding in the healing art, the author submits it to your consideration.

A FEW POINTS FOR CAREFUL CONSIDERATION.

The Importance of a General Knowledge of Domestic Medicine.

Every housewife must necessarily be, to a greater or less extent, a physician in her own family. Not that she can or should altogether usurp the place of the skilled and experienced practitioner, but the ability to recognize diseases and a knowledge of the proper treatment and remedies to apply in the first stages of even serious ailments, will often check their course and prevent what would otherwise result in long and wasting illness, or possibly death.

Further, this knowledge and a supply of simple remedies will enable her to successfully cope with the commoner complaints, to which her household is subject, entirely independent of professional aid, thus saving suffering and many dollars in doctor's bills as well.

Every one appreciates the constant possibility of accident and sudden emergency, to which the whole human race is liable. In nothing is it more imperative than that the mistress of a household be able and prepared to afford prompt relief. If she knows how and has the means to stop the flow of blood from a severe wound, dress the blistered flesh of a serious burn, or ease the pain of a fractured limb, she can often be of as much assistance to ultimate recovery as the surgeon, who later takes the case in hand. Moreover in instances where it is impossible to secure professional aid without delay, sometimes for hours, this skill of the wife and mother may save even life itself.

This necessity for a knowledge of the rudiments of medicine exists in nearly every home, but more particularly is it needful among those whose enviable lot it is to dwell in the country or farming districts, instead of in the city. The latter can secure the aid of a physician, with but a brief delay, at all times, while the farmer's wife must often wait hours before professional assistance can reach her. Then comes the test of her knowledge and foresight, and bitter will be the regret if through want of either, she is unable to give relief to those she loves.

Every woman who has the responsibilities of a family upon her, should arm herself against the time of necessity which will surely come. A little intelligent reading or study and a few common remedies, will sooner or later prove worth their weight in gold.

"Household Remedies."

An extended review of many works similar in plan and scope to this little book, shows that almost without exception they give a great array of "family receipts" for every known disease, to be compounded in the household from common "roots and herbs." While a large proportion of the remedial agents at the command of the medical profession to-day are derived from vegetable rather than mineral sources, from "roots and herbs," yet there are many reasons why the system of recommending these home prepared medicines has wrought untold havoc.

A careful consideration of hundreds of these so-called prescriptions demonstrates that the majority are utterly worthless for the purpose recommended. Many are harmless it is true, while a few are absolutely dangerous in hands unskilled in their proper preparation or administration. Even if harmless, their employment has often resulted in disaster by causing the loss of valuable time. In their incipient stages nearly all of the common diseases can be readily overcome and serious illness prevented. But where one of these useless remedies is relied upon without benefit, until the malady has become firmly seated, the most skilled physician may be unable to cope with it successfully.

The experience of medical men has demonstrated conclusively that few, of even the most beneficial herbs and plants, can be prepared as medicines in the kitchen with satisfactory results. It requires the skill and scientific appliances of the chemist to extract the active principle, the *life* from the plant which is the true remedial agent sought for. This must be secured separated from all impurities and in a form to be readily taken up and assimilated by the system, to afford the desired healing effect. In this the kitchen cannot compete with the laboratory.

It is true, there are many simple remedies easily prepared in every household that are fairly efficacious and these the author has recommended fully with instructions how best to prepare them. But where there is no remedy of this description that can be depended upon, the author has recommended a pharmaceutical preparation that he knows to be the very best obtainable.

There are many manufacturing chemists in this country whose compounds are of the highest degree of excellence, but which are in a form that precludes their being used with safety or success by any but a regular physician. They are not adapted to the laity or for household use.

Therefore, the author has recommended in many instances the various prescriptions and preparations compounded by the *Consumers Drug Company*, knowthat both himself and his readers can rely implicitly on their having the desired effect in every instance. These preparations are the only ones, to the author's knowledge, that are designed especially for household or common use without the supervision of a physician. By following the directions carefully, there is not the slightest danger of evil results, while the effects obtained will be found uniformly satisfactory.

The Family Medicine Chest.

All the knowledge obtainable as to the proper treatment of diseases and emergencies is useless, unless its possessor has also the requisite remedies and appliances. We cannot too strongly impress upon the mind of the reader, the vital necessity of having always at hand, where they can instantly be got at, a few preparations that are needed almost constantly, in every household in the land. In time of need, the delay consequent on having to send to the nearest druggist, may result seriously. One may keep a prescription in the house for months without using it, but when the time arrives that it is required and that instantly, its possession is of incalculable value.

In this connection it may be well to mention the *Household Medicine Cabinet* prepared by the *Consumers Drug Company*. This is a neat case arranged in compact convenient form, stocked with everything that the housewife is liable to require in cases of accident, sudden illness or for the complete treatment of common complaints. It is kept supplied by their local representative and is furnished on terms so liberal, that no family can offord to be without one.

Patent Medicines.

A final word about this one of the greatest modern American evils. The so-called "patent" or proprietory remedies have multiplied so rapidly during recent years and are so generally used, even by those of superior education and intelligence, that they form one of the gravest problems which confronts those who are working for the advancement of the health and happiness of their fellows.

While some of these secret preparations have more or less merit, the majority of them are compounded solely to entice the money of the gullible through clever advertising. If they happen to be harmless the dupe loses only his money. If, as is often the case, they are actually injurious, the victim is robbed of both health and dollars.

It may be taken as a safe rule, never to buy or take a preparation which is not compounded by a reliable firm of manufacturing chemists and such are always willing to furnish their patrons with the formulas of their various remedies, when requested.

HYGIENE.

We have endeavored to impress upon the reader the vital necessity of everyone having some knowledge of disease and its remedy. It is a blessing to be able to succor and relieve suffering, but a thousand fold greater blessing, is the ability to prevent it.

For centuries the law of the Medes and Persians has been the synonym of inflexible rule. But compared with the laws of nature, that of the Persians is as chaff before the winds. Whosoever violates them will as certainly receive meet punishment as time rolls on, it matters not whether that violation be willful or through ignorance. Disease is the penalty dealt out by the outraged Mother Nature to those who have rebelled against her rule.

Ninety-five per cent of the suffering through disease, comes from ignorance of proper living and therefore, can excite only the profoundest pity from enlightened minds. The remainder, which is the result of direct and willful violation on the part of its victims, is the reward of such willfulness.

Progress in the expounding of these natural laws, is making gigantic strides in these glorious years of advancement and the percentage of those who break them ignorantly, is growing steadily smaller. When the happy time arrives, that all American children are obliged to acquire a knowledge of Physiology and Hygiene, as thorough, as well grounded as that now demanded in the rudiments of general education, then and not till then will be seen evidence of the abatement of the awful catalogue of ailments, which the people of this country are groaning under to-day.

It is not our purpose to treat this subject in this connection. It is impossible to do it justice within the scope of this small work, and it is far better to omit it altogether, than to give to a matter so gravely important but the meager outline possible in the few pages at our disposal. Better, by far, utilize them in the endeavor to convince our readers of the importance of a knowledge of correct living and the serious bearing it certainly has on their future happiness.

Secure the best works obtainable on Hygienic regulations and do not read but *study* them. Understand and *believe* them, for they teach the accumulated truths of ages of scientific observation. Their teachings prove themselves. Follow them and you will not lack for convincing evidence.

Then teach your children what you have learned. Begin with the birth of their understanding, constantly adding to their knowledge as their minds expand. Insist on their living rightly. Not only they but future generations of your descendents will bless you. If, when your time comes to take your turn across the dark river, you leave to your children the endowment of perfect health and the knowledge how to retain it, you have given them a far richer heritage than untold millions in gold. It is in this way, and this only, that physical perfection in the human race can ever be attained.

HINTS ON NURSING.

Great as are the resources and achievements of modern medical science and chemistry, they will avail but little in combating disease, without that constant, watchful, intelligent care for the patient and his requirements included under the general term,— nursing.

It is self evident that a professional or trained nurse, who has spent years of study in acquiring every detail of her calling, with unlimited advantages, to be had only in some great hospital, can be of more value in the sick room than the amateur who has little opportunity or practice. But the great majority of families cannot afford the luxury of the professional nurse and the duty, therefore, devolves upon wife, mother or sister.

It is for these, that the following brief hints are intended. While necessarily incomplete, they cover the main points to be observed and we know an intelligent application of them, will afford satisfactory results. They are of a general nature only and brief special instructions as to care, will be found under the head of accessory treatment to different diseases, as described separately.

Light and Air.

Above all things, the sick room should be cheerful and well ventilated, roomy and as pleasant in every way as circumstances will permit.

It is very rarely indeed that a patient requires that the room be darkened and unless this be a necessity, as in some diseases of the eyes or nervous system, the gloom will only have a most depressing effect. In almost every ailment, particularly those of a lingering or wasting nature, such as fevers, it is absolutely essential that the patient be kept cheerful and in good spirits. This is impossible without plenty of Nature's own restorative, sunlight and fresh air.

Even in health, the air of the sleeping room soon becomes noxious, unless constantly renewed and purified and this is greatly aggravated in disease, when the person and discharges of the patient are constantly throwing off poisonous matter which contaminates the atmosphere and surroundings. Ventilation is best had through an adjoining apartment, if convenient, thus protecting the patient from draughts, but when this is impossible open the windows (always from the top) screening the bed and its occupants from direct currents of air. A wood fire place or coal grate affords the very best obtainable ventilation and should always be utilized.

The *Consumers Drug Company's Germ Killer* is a very valuable and necessary adjunct of the sick room. It may be used either by saturating a sponge or cloth and allowing it to evaporate, or spraying the room frequently and liberally with an atomizer. This not only freshens and purifies the atmosphere of the apartment, but destroys the floating poisons of disease and lessens the danger of contagion. No nurse, amateur or professional should ever be without a supply of this most excellent preparation.

The Nurse.

The elements which are absolutely essential to success in caring for the sick, are intelligence, cheerfulness, and presence of mind under all circumstances, good temper, health, strength and neatness.

The good nurse is always bright, alert and absolutely submissive to the directions of the attending physician. No matter what her own theories or ideas may be, she must obey to the letter the instructions of her superior, the doctor.

She must be neat and tidy in her dress, with temper serene and unruffled under the most trying and exasperating demands from her charge. She has everything to do with the mental condition of her patient and oftentimes recovery is entirely dependent on the sufferer being kept bright and sanguine.

She must have the physical strength and endurance to withstand the strain of long vigils without rest, and no woman whose self command is shaken by dangerous crises, or the sight of suffering, to say nothing of bloody operations, is competent to have charge of a sick room.

In substance, a good nurse must be a strong, self reliant, intelligent *womanly* woman.

The Bed.

The sick bed should be roomy, comfortable and placed so as to be easy of access from every side.

Under no circumstance allow a patient to lie on one of those abominations, a feather bed. A firm comfortable mattress is an absolute necessity, preferably of hair, but any cheaper sort will answer when best available.

Light warm blankets are infinitely the best for the sick bed, but the ordinary comforters will answer, if not too heavy. Banish the heavy cotton counterpanes. Their weight is not only often distressing to the patient, without affording compensating warmth, but they serve to retain the poisonous emanations from the person of the invalid.

Pillows should be soft, numerous and of different sizes, when convenient, so as to conform to the various requirements of the patient.

The bedding should be changed frequently, depending on the condition of the patient, and thoroughly aired. It is also best to change and thoroughly air the mattress every few days. In cases of fevers and all infectious or contagious diseases, the sheets should be well washed in water, to which has been added a small quantity of the *C. D. Co's Germ Killer*, to destroy the poisonous matter which is invariably attached to them. Never put sheets or other bedding on the couch of an invalid, without being positive that they are absolutely dry. It is always safest to dry and warm them thoroughly before a fire, before using.

Food.

Brief hints on the proper diet for different diseases, will be found outlined in the treatment for each. The nurse should also be guided by the advice of the attending physician in this very important matter.

The food for an invalid should never be prepared or allowed to stand, either in or near, the room of the patient. Neither should the odors arising from any cooking or eating, be allowed to penetrate to this apartment.

The food intended for the patient should be cooked daintily and served neatly and in its most perfect and palatable form. It should also be given punctually and if circumstances require it, the patient must be induced to eat. In this, coaxing and blandishment, are far more effective than command.

Company.

This is a matter entirely governed by circumstances and can be best regulated by the medical attendant. It is a safe rule however to have too few, rather than too many, visitors. Particularly after a protracted illness, the excitement is apt to have an injurious effect on the weakened convalescent. The patient himself is rarely qualified to judge in this particular and the nurse should enforce her rules, regardless of his protest.

Conversation on business matters should never be permitted when the invalid is at all weak or nervous.

Callers should never be allowed to sit on the bed of the patient, nor should loud talking be permitted. But on the contrary, should it be thought advisable to allow the friends of the sick one to see him, they should be cautioned against showing concern or apprehension at the danger of his condition, and in fact let the conversation be about any bright and cheerful matter, avoiding the subject of the invalid and his ailments, as much as possible.

Cleanliness.

This absolute necessity to health is tenfold more vital in illness. Except in some very few instances, which the physician will quickly determine, the patient must be bathed frequently and regularly, to remove the diseased matter which is constantly thrown off from the system, through the pores of the skin.

Before beginning the operation, the nurse should prepare everything needful, so that it may be as brief and free from unnecessary annoyance or discomfort to the patient, as possible. It is an excellent plan to spread a spare sheet on the bed, underneath the invalid, to protect the bed clothing from dampness, to be removed as soon as the bath is finished.

Bathe only a small portion of the person at a time, using a sponge or soft cloth and drying the surface quickly with a soft towel. Pay especial attention to those parts where the diseased emanations accumulate most quickly, such as arm pits, between the fingers, etc.

Never use anything but pure castile soap. It is not only the cheapest, but the only soap safe for the sick room.

Diet.

The modern physician has come to depend upon this regulation of a patient's food, nearly as much as upon chemical curative agents.

Reliable *general* directions on this important subject are impossible, but the reader will find suggestions as to the proper diet in different diseases mentioned under the description and treatment of each complaint.

ACCIDENTS AND EMERGENCIES.

In case of serious accident or injury, the first and important requisite, is to retain your presence of mind, if you would be of assistance to the sufferer.

Next, is knowing the right thing to do and having the things to do it with. While the following directions, covering nearly all of the emergencies that are liable to occur in the family or neighborhood, are necessarily concise, yet if understood and followed, they are ample to afford relief in serious cases until a surgeon can be had; or the complete cure, without professional assistance, of injuries of the more simple sort. It would be well to memorize the more important, as there is rarely time in emergencies to run and consult a book. Be sure you are right, then go ahead and act quickly.

In the matter of remedies and appliances, we would again mention that handy ally of the housewife, the *C. D. Co's. Household Medicine Cabinet.* This is especially designed to afford prompt help in case of accident and as furnished and kept supplied by them, contains all that is necessary to avert danger, or ease pain. The "Emergency Case" is a most convenient little device and should always be kept fully supplied as directed, with strips of soft, white linen or cotton cloth, (preferably the former) for bandages.

The preparations which have been recommended in the various emergencies, are also to be found in the *cabinet* and are the very best for the purposes indicated, that can be had. If these are in the house (and they should never be absent) always use them in place of any other remedy.

ASPHYXIATION (Suffocation).

Caused by breathing illuminating gas or the gas which escapes from improperly handled coal stoves, also by inhaling the noxious gases of old wells or vats, or the fumes of burning charcoal.

If the patient is partially, or entirely unconcious remove him immediately to the open air, loosening the clothing about the neck and chest to facilitate breathing, sprinkle cold water over the face and head and bathe the temples and brow with *C. D. Co's. Aromatic Ammonia*, allowing the patient to inhale the fumes freely.

If breathing be entirely suspended, endeavor to reestablish it as in instructions for drowning.

Should the case be a serious one, and life apparently almost extinct, a physician should be sent for immediately, as even should the patient revive, the reaction will be severe and demand professional treatment.

BURNS and SCALDS.

These are perhaps the most common injuries which the housewife is called upon to treat, although ordinarily of a comparatively trivial nature. Even small burns or scalds should not be neglected however, as they often develop into serious sores, particularly in persons of scrofulous tendencies.

Without an instant's delay, the burn or scald should be liberally anointed with the *C.D. Co's. Healing Lotion* taking care that the entire surface is well covered. Then apply a thick layer of the *C. D. Co's. Medicated Absorbent Cotton*, wrapping the whole firmly but gently in bandages. After a few hours the *Healing Lotion* will have become absorbed and the burn should be redressed precisely as at first. Always cover it with the *Absorbent Cotton* and bandage so as to exclude the air as much as possible.

Should the burn or scald suppurate, viz—become maturated or discharge pus, bathe it gently in a warm solution of the *C. D. Co's Antiseptine*, in the proportions as directed for this purpose on the label. When the surface is entirely clean and free from foul matter, apply the *Healing Lotion* and dress as before.

The most obstinate and severe burns or scalds will readily and quickly heal under this treatment.

Should the injury be very severe, causing faintness or collapse, the patient should be revived by small doses of brandy or whisky administered frequently and, at the same time, bathing the forehead and temples with *Aromatic Ammonia*.

In cases of small children who are unable to rest or sleep because of the pain, a dose of the *C. D. Co's Soothing Syrup*, will be found to relieve them quickly and induce quiet slumber almost immediately, thus facilitating recovery. This excellent quieting mixture should not be confounded with the many "patent" nostrums bearing a similar name. Almost without exception, these are compounds of opium and highly dangerous. This is the only preparation, to the author's knowledge, which can be given to young children without fear of bad results as it is absolutely as harmless as it is beneficial.

BRUISES.

Insignificant bruises, where only slight discoloration is evident, will rapidly heal themselves without treatment.

If very painful, as bumps on the head, they should be lightly bathed with the *C. D. Co's. Anodyne Liniment*. which will ease the pain immediately. In case the skin is abraded or broken, the *Anodyne Liniment* should never be used, as it is not intended for application to raw surfaces.

Should there be a combination of bruise and slight cut, as from a blow with some blunt instrument, the wound should be bathed in cold water containing a few drops of the *C. D. Co's. Autiseptine* to remove all foreign substances, then a cloth saturated with the clear *Autiseptine* should be bound firmly over the injury. This will be found to have both a cooling and healing affect, very grateful to the patient. Cold applications are used in cases of this sort to check bleeding. If the skin be unbroken they should be applied hot.

Where any of the members have suffered from a severe blow, a careful examination should be made for broken bones, which can readily be detected. (See Fractures). Severe bruises, where there is no abrasion of the skin, should be dressed by applying cloths saturated in a hot solution of equal parts of water and the *C. D. Co's. Anodyne Liniment*, which should be frequently renewed until the pain abates.

FRACTURES or BROKEN BONES.

All cases of broken bones call for the immediate attention of a skilled surgeon and are beyond the province of the housewife, except that she should use every effort to make the patient as easy as possible until the doctor arrives. The physician should always be called without delay, as it is often difficult to reduce the fracture properly after the parts have had time to swell and inflame.

Fractures are of two kinds: simple and compound. In the compound fracture, the end of the bone pierces the flesh, and protrudes so it can be easily seen, while in the simple fracture no exterior wound is visible.

The former is of course apparent, while the simple fracture is often rather difficult to detect, especially after the member has become swollen. It is always accompanied by loss of use of the part, and if the bone has not been heard to snap when broken, it can usually be determined by feeling it carefully to the point where the two ends of the severed part can be felt working against each other.

The injured person should at once be conveyed to the nearest house, and this is best done by carrying him on a stretcher, rather than in a wagon. A door or shutter answers admirably for the purpose. He should be handled with all possible care, to prevent the end of the bone pushing through the flesh, if it is merely a simple fracture, as this would add materially to the injury. If a compound fracture, and accompanied by profuse bleeding, check the hemorrhage as directed for cuts and wounds.

The intense pain of a simple fracture can be greatly relieved by swathing the member with cloths saturated with the *C. D. Co's Anodyne Liniment*, which will instantly relieve the suffering. Should the patient suffer from faintness or prostration, small doses of whiskey or brandy accompanied with bathing the temples and inhaling *Aromatic Ammonia*, will revive him.

In cases of compound fractures, there is no better dressing to be had than the *C. D. Co's Antiseptine*. Use no other.

CUTS and WOUNDS.

These are of frequent occurrence in every household, and while the minor injuries of this description are easily disposed of, it sometimes happens that the flow of blood from a serious wound results in death, because the bystanders have not the knowledge or means to arrest it.

Before proceeding with detailed treatment, we wish to give a word of advice. Never, under any circumstances, apply ointment, liniment, tobacco, court-plaster, sugar, or any harsh remedy, or substance to the raw flesh, particularly to cuts. Not only does this practice needlessly aggravate the inflammation, but it has very often caused the loss of a finger, a leg, or even life itself, through blood poisoning.

After checking the flow of blood, the only dressing admissible is some healing antiseptic lotion, and of these, the best, to the authors knowledge, is the *C.D. Co's Antiseptine*. It is all that is necessary to cure the most severe wounds that the household healer will be called upon to treat.

Unimportant cuts, or abrasions of the skin, call for no further curative measures than stopping the hemmorrhage, by holding the edges of the wound tight together and binding snugly on it a small fold of linen, or bit of *Absorbent Cotton*, soaked with the *Antiseptine*. This dressing should be allowed to remain some time, and when necessary to renew it, the compress may be gently removed by soaking in warm water, should it be attached to the edges of the cut by the dried blood. The bandage and dressing should be kept applied until the edges are completely united.

In even very severe wounds, the services of a physician will not be required if the following directions are carefully followed, unless an artery should be severed, in which case a surgeon should be had at once.

The severance of an artery can always be detected by the blood being of a bright red color, and coming in *spurts or jets*. In this case action must be prompt or the patient will bleed to death in a very brief period. If the wound be in the leg or arm, the member should be bound *above* the injury, viz:—between the cut and the heart, and compressed very tightly. If in the hand or forearm, always apply above the elbow. Take a handkerchief or bandage, knot it around the limb tightly, then thrust a stick under the bandage and twist it until the pressure is so great as to shut off the flow of blood. Do not relax it until the physician arrives.

Should the severed artery be in the head, neck or body, it will be more difficult to manage, but the hemorrhage can usually be arrested, either by pressing the finger hard over the end of the artery, if exposed, or compressing the edges of the wound very tightly together and keeping them so, until the surgeon relieves you.

Should the cut be a severe one, yet without having severed an artery, there may still be very profuse bleeding from the veins. This can be readily stopped by compressing and binding the wound tightly, or using very cold applications.

If the wound is extensive, the next move after the hemorrhage is arrested, is to cleanse it thoroughly of all foreign substances such as dirt, or bits of cloth or wood. If the bleeding be entirely stopped, lukewarm water can be used; but if there is still a slight hemorrhage, the water should be cold. Always put a little of the *Antiseptine* in the water when washing wounds.

Then bind the edges tightly together, by using the strips of *Adhesive Plaster*, which will be found in the Emergency Case expressly for the purpose. Heat them slightly, so they will adhere firmly, and apply them "crossways" of the cut. They should not cover it, but merely bind it together in several places.

Now apply the linen, or absorbent cotton, well saturated with the *Antiseptine*, and bandage firmly. Renew this from day to day, and if the wound suppurates, wash the pus away thoroughly with a strong solution of the *Antiseptine* and renew dressing frequently. You will be astonished at the quickness with which it will unite and heal.

Should the patient become faint from loss of blood, place him flat on his back, bathe the head with cold water and *Aromatic Ammonia* and give mild doses of stimulant.

CHOKING.

This is often a serious matter and sometimes requires quick action to avert death.

If the choking is caused by a morsel of food or foreign substance lodged in the throat, swallowing a bit of bread or some water will often dislodge it.

If this fails, it may be necessary to thrust the obstruction down into the gullet. If it cannot be reached and removed with the finger, a flexible stick like whalebone should be used or a bit of fairly stiff rubber, if obtainable. If harsh this should be covered with silk and oiled.

A common method is to strike the patient sharply in the back which often dislodges the object.

A better plan is given by one medical authority, who states he has often employed it successfully with children. Place the child between your knees, one knee pressed firmly on the back, the other on the stomach. Press one hand on the back between the shoulders and strike the child sharply on the chest.

When any sharp substance as a fish bone, bit of metal, etc., is firmly lodged in the throat, a surgeon should be summoned at once.

CONCUSSION of the BRAIN.

This is caused from a blow on the head or a fall and varies from a slight stun to death.

If at all severe, a physician must be summoned with all haste. Place the patient in a warm bed with the head slightly raised and bathe the brow and temples with *Aromatic Ammonia*. Should he not revive, the doctor on his arrival will apply such treatment as may be necessary and which can only be determined by an experienced physician.

DISLOCATIONS.

Members "out of joint" as the term is, are accidents which are almost certain to confront the domestic healer at some time in her career, and very painful and difficult injuries they often are to manage.

Dislocations of the shoulder, elbow and those of the minor joints of the hand are most common.

They can readily be determined by the malformation of the joint and loss of use of the member, also by the limb usually being either shorter or longer than its fellow.

Except in cases of the smaller joints, as of the fingers and toes, it is always best to call a surgeon immediately, as much depends on the proper reduction of the injury, mistakes frequently resulting in permanent disfigurement. When the surgeon has been sent for, little can be done for the patient but remove him carefully to the house and place the injured limb in the most comfortable position possible. Should he suffer from the shock and show signs of collapse, revive him with small doses of stimulant and bathe his temples with *Aromatic Ammonia.*

If the delay before a physician arrives be great, the intense pain may be greatly alleviated by applying cloths saturated with the *Anodyne Liniment* to the injured joint. After the surgeon has reduced the dislocation he will indicate the further treatment necessary, which will be determined by circumstances.

In places far removed from civilization, where a physician cannot be had without many hours delay, a dislocation of the shoulder if it be upward, forward or backward can often be reduced by placing the foot (from which the boot has been removed), under the arm pit of the patient and pulling the arm steadily and firmly, while an assistant endeavors to work the joint into its socket with his hands. Care should be taken in doing this not to rupture the small muscle which crosses at the *front* of the hollow of the arm pit. When the shoulder is in place, bind the arm securely to the body and swathe the injured part with cloths saturated with *Anodyne Liniment,* which will relieve the pain and drive away the soreness.

Small dislocations can be easily reduced, by pulling the member firmly and at the same time working the joint back into its place with the other hand. Dress with the *Anodyne Liniment* and the pain and soreness will soon pass away.

DROWNING.

Unless a person has been in the water some time and is without question entirely dead, never give up without using every means of recusitation, even if there is not the slightest life apparent

As soon as the patient is taken from the water, the first move must be to free the throat and bronchial tubes from water, and reestablish breathing. Turn him on his face and with the finger slightly crooked, depress and pull out the tongue, allowing free egress to the water, which will run out readily if the head be depressed. This is a much more effective and humane process than suspending the patient by his heels, or rolling him over a barrel.

Cleanse the mouth and nostrils thoroughly. Strip the patient of his wet clothes and envelope him in warm blankets. Endeavor to reestablish warmth and circulation by chafing the extremities, and rubbing the whole surface of the body briskly.

If the breathing has entirely ceased, it must be reestablished by artificial means. One method is to breathe into the mouth of the patient until the lungs are inflated, then expel the air by pressure on the sides of the chest. This should be done about fifteen times a minute.

Another plan and one that is considered very effective is as follows: Place the patient on his back on a firm hard surface, the body slightly inclined from the upper portion downward, with the head and shoulders supported by a small pillow or folded coat, and see that the tongue is kept drawn forward and the throat open. Stand at the patient's head, reach forward and grasp both his arms at the elbows and draw them directly upward, until they meet above his head. Hold them a second or two and then return to his sides, pressing them firmly against the sides of the chest. If an assistant compress the lower part of the ribs and the diaphragm at the same time the arms are pressed against the sides, the operation will be facilitated.

Never allow the patient to inhale ammonia or other restoratives until breathing is firmly established, when small doses of stimulants can be given to promote recovery.

Always persevere in efforts to recusitate a drowned person for at least an hour, for many apparently hopeless cases can be saved by patience and hard work.

ACCIDENTS to the EYE.

The most frequent of these is the introduction of foreign substances or particles. If under one of the lids, fold it back, asking the patient to look downward if the upper lid and upward if the lower, when the offending object can be easily seen and removed with a small soft brush or the point of a lead pencil.

When sharp objects, as bits of steel, become imbedded in the ball of the eye, a surgeon should be called, as the inexperienced should never risk using the knife or other sharp pointed instrument.

Blowing the nose sharply will sometimes remove intruding eyelashes or other irritants. Always avoid rubbing the eyes when irritated, as it only adds to the difficulty.

Should the eyes become inflamed, either from the irritation of foreign matter or other causes, there is no better remedy than the *C. D. Co's.* prescription *No. 37* which allays the inflammation and has a generally soothing and healing effect.

Black or contused eyes, resulting from sudden blows, can be relieved by applying hot cloths to the part, while the soreness and pain will be dissipated by rubbing on the *Anodyne Liniment*, taking care not to get it into the eye itself.

EARACHE.

If caused by any foreign substance having become

lodged in the ear, its removal will always relieve the pain. This should be accomplished with small forceps or the introduction of tepid water with a syringe.

If the earache arises from inflammation caused b cold, as is usually the case, heat a brick or stone, wrap it in a wet cloth and thoroughly steam and sweat the part. Then saturate a bit of *Absorbent Cotton* in slightly diluted *Anodyne Liniment*, place it tightly in the orifice and the trouble will soon cease.

FAINTING.

The first and vitally important move in cases of fainting, is to place the patient flat on her back without any pillow or support for the head. Quickly unfasten the clothing about her throat and also loosen the corsets or other tight garments.

Then bathe the brow and temples liberally with *Aromatic Ammonia* allowing her to inhale it plentifully, and sprinkle her face with cold water.

Should she be weak on reviving, administer small doses of brandy or whiskey.

FREEZING.

It is very essential in cases of either severe freezing and utter prostration from exposure to the cold, or simple frost bites, that the patient be kept from sudden contact with the fire or warmth. It not only aggravates suffering, but is occasionally positively dangerous.

Rub the affected parts with snow or immerse them in cold water until circulation is reestablished. Should the hands or feet become sore, treat them as directed for chilblains.

In case a person is prostrated from exposure, rub with snow until circulation is established and revive as directed for collapse from other accidents or injuries.

GRIPING.

The intense grinding pains through the stomach and bowels, are almost invariably caused by errors of diet and usually precede or accompany colic and diarrhœa.

Give the patient frequent doses of the *C. D. Co's Cholera Cure* and apply cloths saturated with a mixture of very hot water and *Anodyne Liniment*, over the affected part. These cloths should be kept as hot as can be borne and removed frequently until the pain ceases.

GUN SHOT WOUNDS.

These injuries always call for the prompt attendance of a skillful surgeon and one should be summoned immediately.

Pending his arrival, should the patient suffer from faintness or collapse, he should be revived and sustained as directed for other accidents.

PROSTRATION from HEAT—SUNSTROKE.

If the person is only slightly prostrated, suffering from giddiness, weakness, etc., cold applications to the head combined with liberal bathing of the temples with *Aromatic Ammonia* will usually relieve him. It is always best that he be kept quiet, in a cool and darkened room, for at least several hours after the unpleasant sensations have passed off, in order to insure against recurrence of the difficulty.

If the symptoms are those of severe sunstroke, accompanied by delirium and raving, a physician should be summoned immediately. Until his arrival, efforts should be made to effect relief as above.

HYDROPHOBIA.

Fortunately, this horrible disease is very rare, as it is one that is almost beyond the physician's power to subdue. The treatment by inoculation, discovered by M. Pasteur, the famous French scientist, is the only one that is known to be successful. Cauterizing the wound will often destroy the deadly effect of the poison, and should always be done by a physician. It is always safest to cauterize the bites of all dogs, whether suspected of rabies or not.

LOCK-JAW.

This distressing malady usually arises from slight wounds that have not been properly treated and is characterized by the jaws of the patient setting, or locking together so tightly that it is almost impossible to open them. It is sometimes accompanied by more or less violent spasms, and very often results in death.

At the first symptoms of rigidity in the muscles of the jaw or neck, a physician should be summoned, as none but an experienced expert can successfully treat *tetanus*.

BLEEDING from the NOSE.

Except when very profuse and long continued, hemorrhage from the nose is of little importance, and often in persons of a very full habit and apoplectic tendencies, it may have a positively beneficial effect.

It can almost invariably be arrested, when desired, by snuffing cold water up the nostrils, applying cold cloths or ice to the back of the head or neck, or holding the arms high above the head.

POISONING from IVY or other NOXIOUS PLANTS.

The parts should at first be covered with common baking soda made into a thick paste with water and after this is removed, kept thoroughly anointed with the *C. D. Co's Rose Cream*, which will rapidly allay the irritation and the pain. Should the case be severe and rather obstinate, the *C. D. Co's Sarsaparilla Resolvent* should be given regularly as directed on the label. The *Sarsaparilla* will act quickly in eliminating the poison from the system.

This treatment, if persevered in, will overcome the worst cases of poisoning, either from ivy, poison oak or poison sumach, without the aid of a physician.

SPRAINS and STRAINS.

Few accidents are more often brought to the attention of the housewife, than these, and it is very rarely that she will require professional aid in quickly and completely curing them, if she is supplied with that invaluable resource, a bottle of *C. D. Co.'s Anodyne Liniment.*

If the sprain be of the ankle, knee, elbow, or wrist, immediately remove the clothing from about the part, and apply the *Anodyne Liniment* with the palm of the hand, rubbing the joint rapidly and firmly on all sides and *using an upward motion.* As the liniment becomes absorbed, keep adding more and continue the application and brisk rubbing for at least half an hour. Then swathe the parts in cloths saturated with the liniment and keep the patient perfectly quiet, not allowing him to use the injured member in any manner.

Repeat the rubbing process several times daily alway using the *Anodyne Liniment.* At first it may cause the patient some added pain but persevere notwithstanding his protests and you will be rewarded by a complete and early recovery.

Strains of the back, such as those caused by undue lifting, falls, etc., are matters of serious importance and should be promptly attended to. If the pain be very sharp, use *Anodyne Liniment* with rubbing as directed above. After which apply one of the *C. D. Co's. Penetrating Porous Plasters*, which are designed especially for this purpose. You will find that their beneficent action will quickly reach the sore spot and draw to the surface the humors which would, if undisturbed, cause extended if not permanent weakness and discomfort.

SNAKE BITES.

The most common of the venemous reptilia in the United States, and at the same time one of the most deadly is the rattle-snake. Fortunately this snake is easily identified and there can be no question as to the poisonous character of the bite, therefore active measures for relief will not be delayed.

When a person is bitten by a poisonous snake on one of the extremities, the first care should be to prevent the circulation of the virus through the system. This can be effected by binding the limb tightly between the bite and the heart, precisely as for a ruptured artery.

The poison should be extracted from the wound, if possible, either by sucking it with the mouth (care should be taken that there are no breaks or abrasions in the skin of the mouth,) or it has been recommended by some writers, that a chicken or small animal be quickly killed and a portion of its body applied to the bite while the flesh is still warm and palpitating. Ammonia has also been applied to the wound with satisfactory results.

In the mean time, the patient should be given frequent and large doses of spirits, either whiskey, brandy or rum until he became thoroughly intoxicated.

This treatment has been employed many times with uniformly satisfactory effect and is probably the only one that can be relied upon.

STINGS of POISONOUS INSECTS.

The stings of bees, wasps, and hornets as well as those of gnats and mosquitos, can be relieved by bathing the part frequently in *Aromatic Ammonia* or strong salt and water.

POISONS AND THEIR ANTIDOTES.

In cases of poisoning, always remember that there must not be an instant's delay, if one would save the life of the patient. Once let the poison get into the system and no antidote will be effective. It must be removed or its effect counteracted almost the moment it is swallowed. It is also advisable to send for a physician immediately, even if the proper antidote has been applied and afforded some relief.

In the following instructions, the author has confined himself to those varieties of poison which are most liable to be accidently taken in the household, and has also prescribed only those antidotes which can always be had in every home, believing that these omissions will not detract from the practical value, but add to the clearness of his treatment of the subject.

Emetics.

Vomiting is the only remedy for the majority of poisons and to save unnecessary repetition and confusion, we will outline the easiest and most effective methods of causing the stomach to evacuate its contents, before considering the different poisons separately. The first and best is to mix a teaspoonful of the *C. D. Co's. Medicinal Mustard* in a glass of warm water, which the patient should drink quickly. If the stomach does not instantly respond cause him to drink two or three glasses of tepid (not hot) water in rapid succession, then tickle the throat with the finger or a feather. An adult's dose of the *C. D. Co's. Ipecana* is a good emetic free from irritating features. Either of these is almost certain to produce vomiting, but if not quick enough use them all. Of course a physician's stomach pump is best of all.

Arsenic.

This is a common poison, being often found in coloring matter, paints, Paris green, etc. Cause vomiting at once and then give plenty of warm milk, sweet oil or the white of eggs.

Acids.

Give freely of strong soap and water, wood ashes mixed with milk, or lime water, to counteract the acid.

Ammonia.

This is a powerful alkali and should be neutralized by giving the patient an acid as a counteractant. Vinegar or lemon juice in teaspoonful doses until relieved, will answer.

Corrosive Sublimate.

This deadly drug enters largely into many rat and insect poisons and may therefore come within the housewife's experience. Mix the whites of a dozen eggs in a pint of water and give to the patient until the stomach will hold no more. Flour mixed with water is also a good remedy but the stomach pump is better.

Opium, Laudanum, Morphine.

When suffering from an overdose of opium or any of its preparations, the patient is determined to sleep and evinces great reluctance to get up and move about. The mustard emetic should be given immediately and repeated until the stomach responds. Follow this with large doses of the strongest coffee that can be prepared and repeat frequently. Keep the patient constantly in motion without rest until the effects of the drug have worn off.

Phosphorus. Matches.

If a child has eaten the heads from matches and shows evidences of poisoning, vomiting should at once be excited, which follow with doses of flax-seed tea or any mucilaginous drink. Carefully avoid all fats or oils.

Strychnine.

This is one of the deadliest poisons known and even prompt action will not always save the patient's life.

Give freely of any fat, as lard or sweet oil and excite the stomach to reject it by tickling the throat. Repeat frequently and have a physician instantly.

Salt Petre, Lye, Potash.

Give large doses of sweet oil, castor oil or lard, to counteract the alkaline properties of the poison. Follow with mucilaginous drinks. Do not irritate the stomach with mustard emetics.

Turpentine.

Prompt vomiting, followed with whites of eggs, milk, or flour and water.

COMMOM DISEASES AND THEIR TREATMENT.

We cannot impress upon the minds of our readers too strongly, the important fact that a great many, if not all of the commoner diseases to which humanity is heir, can be prevented by correct living and sanitation, or can, if taken in their earliest stages, be dissipated or broken up by prompt treatment.

As regards the sanitary arrangements and cleanliness of the household and its surroundings, modern science has demonstrated that a very large proportion of ailments are communicated to the human system through germs or bacilli, which are either constantly floating through the atmosphere and are absorbed through the lungs, or, as with the typhoid and infectious microbes of similar character, they reach their victims through the medium of water and food.

Defective sewers, cess pools, etc., are great swarming places of these insidious and deadly enemies of health and if the well-being of your family is of value to you, see that these are properly located and constructed and above all, keep them thoroughly clean and disinfected.

No matter how perfect the arrangement of these receptacles for refuse matter, the germs of disease are bound to breed and multiply in them, unless the warfare against this enemy be constant and unflinching.

It has been fully proven that only with the most powerful products of chemistry, can they be destroyed, and no household in the land should be without a constant supply of strong disinfectants with which to keep down these multitudious foes of health.

One should as complacently allow the larder to remain empty as to be without a well filled bottle of disinfecting liquid.

The most efficacious of these, to the author's knowledge, is the *C. D. Co's Germ Killer* and it is supplied at a price which places it within the reach of everyone. A constant and liberal use of this invaluable preparation, will prevent sickness that would cost the family a thousand times the expense of prevention in money expended in doctor's bills, to say nothing of the suffering and anguish of disease and death.

A final word as to prompt action and we will pass on to a detailed consideration of the different diseases.

There are few, if any, serious ailments which do not begin with what seems merely slight indisposition of the patient. Never should these apparently unimportant complaints be passed over without attention from the wife or mother. This is the time to act and act quickly, if you would ward off serious consequences.

A slight cold quickly develops into pneumonia, or even consumption, if not checked.

Simple diarrhœa is possibly only the forerunner of inflammation of the bowels, or in these days of apprehension, the deadly Asiatic cholera itself.

The unimportant attack of biliousness may mean jaundice, or a life of constant misery through the diseased liver, if not warded off.

That dull headache and feeling of lassitude is not unlikely to prove the warning of the coming of a wasting fever, which will smother life in its insidious folds.

All these can be combated with success before their grasp is firm on their victim, but after their hold is fixed, all the skill of science may not avail.

So we say to the woman with whom this responsibility rests, be alert and quick to act. At the first symptom of indisposition in your family, use every effort to determine the disease, then administer the treatment necessary to destroy it.

Indifference or carelessness may mean life-long sorrow.

ASTHMA.

Causes.

Irritation of the nerves of the organs of respiration, resulting usually from deranged digestion or an impure condition of the blood.

This disease is very liable to recurrence in persons who have once suffered from its attacks and it is generally acknowledged to be transmitted by inheritance.

Symptoms.

Paroxysms or spasms of difficult breathing, attended with a wheezing sound and a great sense of pressure or constriction across the chest.

The attacks are most common during the night, although they are liable to occur at any time and are always aggravated by improprieties of eating.

Some patients may suffer from a nearly continuous dificulty in respiration, while others appear and feel entirely well, except during the periodical spasms.

Remedies.

A change of locality is of unquestioned benefit in nearly all cases, but no definite rules for this can be laid down, as the climate which would greatly increase the disease in one patient, will eradicate it in another.

The available remedy which affords the most pronounced relief, is the use of the *C. D. Co's Asthma Pastilles* combined with the *C. D. Co's Sarsaparilla Resolvent.*

The pastilles or wafers when burned and the patient allowed to inhale the fumes, will afford immediate relief in the severest paroxysms, while if burned in the bedroom regularly every night as a precautionary measure, they will ward off attacks of the disease and insure healthy undisturbed slumber. No sufferer from this distressing malady should be without this certain means of relief.

The *Sarsaparilla Resolvent* strikes at the very seat and cause of the ailment, purifying the blood and invigorating the organs of digestion, thus eliminating the disease entirely from the system.

Asthmatic subjects should take it regularly, persevering even though the beneficial results of the first few doses are not apparent, remembering that a disease so deep seated and obstinate as this, will not yield without prolonged and steady treatment.

The reward of such perserverance is sure to come.

Accessory Treatment.

Persons subject to asthma should carefully avoid

eating hearty suppers, and in fact commit no indiscretion of diet. Over-exertion, as running, etc., should also be shunned.

Smoking tobacco, either in a pipe or cigars, is of unquestioned value as apreventive, habitual smokers being less liable to attack.

AGUE—(Chills and Fever.)

There are very few sections of the country in which this disease has not been epidemic at some period, but almost invariably it has been found to prevail most largely in newly settled localities. It seems to gradually disappear as civilization progresses and in the older states is to be found only occasionally, and that invariably in low, marshy surroundings.

Its direct malarial origin is admitted by all authorities.

Symptoms.

Alternate, periodic attacks of burning fever, intermitted with chills and great depression of both the physical and mental system.

Its character is so marked and distinctive, that there need be no question of the real identity of the disease, as soon as its first symptoms are developed.

Remedies.

Like all malarious diseases, this is a most obstinate complaint to overcome and will only yield to special and thorough treatment.

The most successful remedies seem to be those devised by old practitioners who have had long experience in treating this malady in all its phases and forms. One of the best, in fact the most reliable of these old standard prescriptions, has been secured by the *Consumers Drug Co.*, and is prepared by them in a most palatable and convenient form.

It is a veritable specific for this harrassing ailment and those residing in malarious districts and subject to intermittent fever, should never be without a supply of the *C. D. Co's Ague Cure*. Each bottle is accompanied with full and complete instructions for accessory treatment, etc.

BILIOUSNESS.

That thoroughly disordered condition of the liver and digestive apparatus, whose milder forms are generally known under the comprehensive term of biliousness or bilious attacks, are among the most common complaints which call for household treatment. Not only does the real suffering with which they are accompanied call for early relief, but they indicate disturbances, which if neglected and allowed to become firmly seated, are liable to develop into serious and often dangerous illness.

Causes.

As these attacks come from irregularity in the action of the liver and organs of digestion, they necessarily have their origin either in constitutional weakness of these important portions of the human machinery, or are brought on by indiscretions in diet.

Many persons are subject to periodic visitations of this complaint, often at almost regular intervals and such must not only exercise great care in the regulation of their eating, but should also take such measures of precaution as will tend to avert the attack.

Symptoms.

As in all diseases which are of so comprehensive a nature, the symptoms are distinct and vary greatly in different cases. Constipation is an almost certain accompaniment and until relieved, the trouble cannot be abated. Many patients are affected with nausea and vomiting and another prominent symptom is a sallowness of the complexion, with a pronounced yellow tinge to the whites of the eyes.

But there is no more common or painful form of biliousness than the malady known as *Sick Headache*. Those who have either felt or witnessed the suffering of this distressing illness, need no directions to enable them to identify it.

Remedies.

Obviously the only remedy is to correct the functional derangement of the organs involved, invigorating and toning them up to the standard required by nature to perform their allotted work. When these portions of the digestive apparatus are not strong enough to throw off the refuse matter constantly accumulating in the system, it soon poisons the tissues and unless quickly removed, nature protests most forcibly.

To stimulate and quicken this action, no product of medical study and chemist's art can equal the famous *C. D. Co's Liver Pills*. Their action is quick, mild and effectual and they form a sovereign specific for all the ills of the digestive system. In acute cases, a dose of three or four will afford rapid and certain relief, while persons of bilious temperament can effectually ward off the complaint by occasionally taking one of the little pellets before retiring. The most obstinate and confirmed cases are compelled to yield to their gentle but inflexible control.

Ladies whose complexion is sallow or muddy will also find them of infinite benefit.

Accessory Treatment.

Persons subject to attacks of biliousness, should use great care in the selection of their food and avoid excessive eating or drinking, as well as late suppers, etc. Coffee, when used too frequently, has been found to have an unfavorable influence on those of bilious temperament.

BOILS.

These annoying pests are so well known to everyone, that extended mention of their characteristics is unnecessary.

Should it be thought advisable to disperse or "scatter" them, it can usually be done by frequently rubbing the spot with the *C. D. Co's Anodyne Liniment* as soon as the first soreness is felt. It is better, however, to bring them rapidly to a head by means of poultices, which should be continued until suppuration has ceased and the inflammation is subsiding.

But the best treatment of all, is to remove the cause and renovate the impoverished and impure blood, which makes known its condition through these irruptions. No preparation excels or equals the *C. D. Co's Sarsaparilla Resolvent* as a blood purifier, its action being to tone up the entire system and eliminate from it, by natural channels, those poisonous elements, which would otherwise be forced to seek egress through boils, carbuncles, etc.

These warnings of kindly nature should not pass unheeded, or serious illness may ensue. They are given us for a purpose and those who value health will obey the signal, and immediately pursue a course of thorough alterative treatment. Remember, the *C. D. Co's Sarsaparilla Resolvent* is a reliable prescription which will promptly and efficiently perform its work, and it should never be confounded with the widely advertised nostrums, which are too often prepared from cheap and impotent materials, merely to reduce the cost of manufacture.

BUNIONS.

Enlargement of the tissues of the great toe, forming apparent deformity of the joint.

As they are caused solely by improperly fitting boots or shoes, the only remedy is to correct this fault and wear loose and comfortable foot gear. The pain may be assuaged by wrapping the swollen joint in bandages saturated with *AnodyneLiniment*, or applying soft relaxing poultices.

BACK ACHE.

This may arise from any of several causes and in any case, it should never be neglected.

As the result of a sprain, its acuteness should pass off in a short time if remedies are applied, and if caused by a slight touch of rheumatism, the same curative agent, viz:—the *C. D. Co's Penetrating Porous Plasters*, will afford relief more quickly than any other means, to the author's knowledge.

But the pain and sense of weakness through the small of the back, may be a premonitory symptom of a serious affection of the kidneys and as a momentous danger signal, should receive careful attention.

Should the *Penetrating Plaster* not have the desired effect in a brief period, other confirming indications of kidney trouble should be looked for.

The reader is referred to the chapters on Bright's disease, diabetes, etc., which give in detail the conditions which indicate a diseased condition of these vital organs.

Should even a very few of them be visible, not a moment's time should be lost, in inaugurating a most vigorous treatment, if one would rout that demon Bright's disease.

BRIGHT'S DISEASE of the KIDNEYS.

The ravages of this terrible malady seem to have been on the constant increase during recent years, until it may fairly be termed a veritable plague of the American people. Each year its victims are numbered by thousands and with every recurring season, the proportion is greater.

Its real origin is enveloped in doubt and the opinions of medical writers of the highest standing, on this matter, would fill many of these pages, if reproduced.

Moreover, it would be an array of conflicting theories and contradictions in the highest degree confusing. The cause of its development, however, seems to be traceable directly to exposure, intemperance, or as the sequel of some fever.

Symptoms.

The proofs of its malignant presence are, with one exception, nearly as many and varied as the opinions regarding its cause. Different patients have an entirely different set of symptoms, with the exception of that almost infallible testimony, the presence of albumen in the urine. This deposit is never absent in a case of Bright's disease.

Dropsy and swelling of the extremities is a frequent accompaniment. The complexion is pale and puffy and there are often periodic attacks of nausea and vomiting.

The urine is scanty, of a more or less dark or smoky color and the patient has an almost constant desire to void it. On standing, it deposits a large quantity of thick, dark sediment.

Remedy.

Whether it is possible to effect a complete cure of thoroughly developed Bright's disease, is a matter of decided question. But that it can be eradicated in its early stages, has been demonstrated, and with proper treatment, even firmly-established cases will so far yield, that the patient will not only live much longer, but his measure of life will be much happier.

The *Consumers Drug Co.*, compounds a system of treatment, the discovery of an eminent specialist, which has been used with the greatest success, both in the cure of the disease in the incipient stages and the relief of more chronic cases.

It is not claimed that the preparation is an absolute cure for every case of this obstinate complaint, for there are many in which complete cure is impossible, but that it will entirely eradicate it in many instances and afford relief in all, has been proven.

A card to their nearest representative, will bring you a comprehensive treatise on the disease, giving full particulars regarding the remedy and also accessory treatment and diet, matters of vital importance.

CONSUMPTION.

We do not believe a very extended description of the causes and indications of this dread disease is necessary. Who has not seen numberless examples of the insidiousness and the merciless grip with which it strangles its victims? That it is one of that class known as germ diseases is now conceded by all authorities. Everyone is to a certain extent subject to its contagion, but only those whose organs of respiration are inherently weak, or whose general system is debilitated, need succumb. The strong and vigorous are invulnerable to its assaults. The bacilli find no lodgement and are innocuous. But at the slightest laxity of this resistance, the germs at once insinuate themselves and then nothing will entirely remove them.

Consumption is inherited, in that the constitutional weakness which renders easy the lodgement of the microbes, descends from one generation to another.

Should this tendency once have shown itself in the

family line, only unfaltering vigilance will prevent its reappearance.

Remedies.

Slight colds are a very frequent cause of weakening the respiratory organs, to an extent that permits the entrance and consequent rapid progress of consumption. Never neglect a cough or cold however unimportant it may seem. Always check it with the first symptom. For this nothing excels in effectiveness the *C. D. Co's Cough Balsam.*

And this calls up a word of caution.

Beware of the host of cough remedies which are advertised as cures for consumption. The *C. D. Co's Cough Balsam*, the best preparation for coughs and colds ever made, will no more *cure* consumption than a pound of sugar will sweeten the Atlantic Ocean. Therefore the so-called cough-consumption specifics are self evident frauds and should be shunned accordingly.

A good cough mixture is of untold benefit in *preventing* a serious affection of the tissues which would doubtless lead to tuberculosis, but it would not have the slightest remedial effect after consumption is firmly established.

The only remedies which have proven efficacious in checking the progress of the disease, are change of climate and cod liver oil.

Unless too far advanced, consumptives can almost invariably secure relief by removing to a dry bracing climate, such as that of the Colorado mountains, or certain of the pine-wooded hill sections of the south.

But the most available remedy for all classes of invalids is cod liver oil. The nutritive properties of this product are most marked, and it is to this and the ease with which it is digested, by even the most feeble, that it owes its wonderful curative powers.

Consumption is essentially a wasting disease. Counteract this waste and its progress ceases. Build up the tissues and increase the general vitality of the system and unless too firmly fixed, the disease will have to give ground and eventually be thrown off.

The pure cod liver oil in its natural state is very unpalatable to everyone and nauseating to the sensitive stomach of the invalid. To remedy this a modification of its form has been devised, often combining the oil with other curative and stimulating agents and these emulsions, as they are termed, have entirely superseded the natural oil as a remedial preparation. There are several brands of these compounds of excellent quality, but none are better than the *C. D. Co's Emulsion of Cod Liver Oil*.

Its basis is the purest quality of refined Norwegian cod liver oil that can be obtained and the process which gives it the distinctive delicate flavor, so acceptable to the weakest stomach, transforming a nauseous oil into a really pleasant article of food, is a purely mechanical one that does not in the least impair, but on the contrary adds to, its curative value.

It should be used in preference to all others, as being both the best and the cheapest that can be had.

When requested, the *C. D. Co.* or their nearest representative, will furnish, free of charge, a little pamphlet on consumption, which gives full instructions as to accessory treatment, etc., therefore the author will not dwell further on the subject in this place.

CHOLERA MORBUS.

In its milder forms this complaint is very common, especially during the later summer and autumn months.

Its exciting causes are various, although an attack is usually directly traceable to either exposure or errors in diet.

While ordinarily yielding to simple treatment without difficulty, the possibilities of this very painful ailment should not be treated lightly. Fatal terminations of an attack of cholera morbus are very frequent and there are few diseases the extremely severe forms of which, cause death more quickly.

Symptoms.

Intense griping pains through the region of the bowels and stomach, accompanied by more or less diarrhœa and often nausea and vomiting. As the disease progresses, the extremities become cold or cramped and covered with clammy perspiration.

The pulse is at first quick and irregular, afterward sinking, as collapse comes on, until almost imperceptible. Unless relief is afforded, death will very often follow.

Remedies.

The *C. D. Co's Cholera Cure* is an almost infallible remedy for this and its kindred complaints. Given frequently it will unaided subdue very severe attacks.

When the patient is suffering intense pain, however, local applications should be added to the internal treatment. Apply either cloths saturated with equal parts of water and the *C. D. Co's Anodyne Liniment*, very hot, or a mustard poultice directly over the seat of the pain.

Some persons are very subject to attacks of this dangerous complaint and such should always avoid both over-eating, indigestible food or unripe fruit and exercise great care in regard to undue exposure.

COUGHS and COLDS—BRONCHITIS.

These are the almost constant care of the housewife, especially during the winter season. But, armed with a bottle of that peerless specific, the *C. D. Co's. Cough Balsam*, she need have no fear of them.

While a repetition of what the writer has stated many times in other chapters, the caution cannot be too strongly emphasized. Never neglect what may seem but trivial illness. The apparently unimportant cold may without warning develop into pneumonia, or even consumption, if left to take its course.

Catarrh in its chronic form will not be treated in this connection, nor will the more serious affections of the throat and lungs, all of which will be found under their proper heads.

We are referring now to simple colds and their usual consequence, bronchitis.

Remedies.

The *C. D. Co's Cough Balsam* should be given with the first symptom and it will invariably soothe and heal the inflamed and irritated tissues of the throat or bronchial tubes.

As accessory treatment, especially when the disease

is located in the nasal ducts, the patient should be put to bed and made to perspire freely. Lemonade made with hot water, mixed with a small amount of brandy, will prove effective. Soaking the feet in hot water and mustard is also good.

If there is considerable pain across the chest and through the bronchial tubes, a poultice made of the *C. D. Co's Medicinal Mustard* will prove highly beneficial.

After any of these and in fact at all times during the progress of even a slight cold, the patient should not be allowed to expose himself, either to draughts or in the outer air.

CHAFING.

This very annoying difficulty can usually be remedied by frequently bathing with a solution of soft water and common alum, applied three or four times a day. It should be used cold and the parts afterwards thoroughly dried with a soft towel.

Powdered alum or fuller's earth, dusted on the parts will often heal the abrasions.

CORNS.

Only those who have suffered from them, (and this, we believe, includes the major portion of civilized humanity), can appreciate the amount of inconvenience, not to say genuine anguish, which can be caused by one little innocent-appearing corn.

Its cause is known to everyone, and until the improperly fitting boot or shoe is superseded by one of correct size and shape, no remedy will be of the slightest effect in removing the pest.

Then drop a card to the nearest representative of the C. D. Co., for a package of *Cornicide* and the neat little treatise on the proper care of the feet, which accompanies it. The twenty-five cents which it will cost, will bring you in a thousand times that amount of peace and immunity from annoyance.

What has cured thousands will relieve you.

CHOLERA.

The reappearance, in some parts of Europe during the season of 1892, of this grim destroyer, which has at different periods of the world's history devastated nearly every civilized land, leaving behind it a blackened swath of dead and dying, has caused widespread interest and apprehension among all classes.

It has been many years since the United States has suffered from a severe visitation of this terrible epidemic, but the firm hold which the plague seems to have secured on Continental Europe and the obstinacy with which it resists eradicating measures, indicates that it will be hardly possible for this country to escape during the coming year.

Consequently, it behooves everyone, no matter what his occupation or surroundings, to be thoroughly conversant with the characteristics of this most deadly of all known diseases. It is no respecter of persons and the farmer in the healthy uplands or the millionaire in his palace, is as liable to its grisly embrace, as the ragged beggar in the crowded, reeking slums of a great city.

Its descent is so swift and awful, that unless armed at all points with the teachings and defenses of science, resistance is impossible.

Causes.

It has been demonstrated beyond any possibility of doubt, that cholera is a germ disease. Like all of this terrible class, it is highly infectious.

The reason of its periodic appearance, alternating with years of almost absolute freedom from its sway, has so far baffled scientific observers. But that at times everything is swarming with these deadly bacilli, while again there are long periods in which they are totally absent, the history of the disease since its first recorded appearance in Madras in the Indian Empire in 1769, has proven conclusively.

Our space will not permit even a brief history of the epidemic, nor an account of the investigations which have led up to the present accepted theory of the origin of the disease. Suffice to say, that the recent visitations in Hamburg and other European cities have proven its correctness.

Preventive Measures.

Inasmuch as the cholera germ, in common with all others of its disease spreading fellows, inhabits and multiplies in all receptacles for refuse and waste places, one of the most certain means of preventing the epidemic is perfect sanitation. Keep every possible lurking and breeding place for the fatal microbes clean and free from decaying matter. Further, make them absolutely untenable to the bacilli by the liberal and constant use of disinfectants. Use the *C. D. Co's Germ Killer* morning, noon and night and the foul brood must depart. This powerful combination of chemicals is certain death to the entire family of disease germs and as a measure of safety, should never be absent from the household.

Great care should be used in the selection of food and water, for experience shows that the poisonous atoms can only secure lodgement in the system through the stomach. They may be breathed with impunity, but once absorbed with the food or drink and they seek lodgement in the intestinal canal from which they quickly sap the life of their victim, unless destroyed. Therefore all water should be thoroughly boiled before drinking, and only plain nutritious food should be partaken of, avoiding unripe fruit and every substance which might tend to derangement of the digestive organs.

Symptoms.

Cholera begins almost invariably with simple diarrhœa which if unchecked is rapidly followed by vomiting and all the severer phases of the disease.

This purging is at first characterized by very watery discharges, which later assume the peculiar character known in true Asiatic cholera as "rice water." With these latter almost always come cramps, in which the patient writhes in terrible agony. The countenance assumes a leaden hue, seeming to shrink and the eyes have a stony, staring look.

Then follows the last stage, that of collapse, when periods of wild delirium are intermitted with times of complete exhaustion and partial or absolute insensibility.

This is ordinarily of short duration and when the disease has been allowed to advance unchecked to this point, there is little to hope for but death.

Remedies.

Whether there is any remedial agent that is uniformly effective in combating true or Asiatic cholera, when it has firmly closed its grip on a patient, is rather doubtful.

But the disease can be checked and eradicated in its first stages, providing the right means are employed and no better or more effective agent is available than the *C. D. Co's Cholera Cure*.

This prescription was originally the discovery of a prominent physician in New York, who used it with the greatest success during the epidemics of 1848 and 1853.

While it is a specific for all purposes requiring the use of an astringent, its power in battling with the cholera germs is peculiarly marked and we know its prompt use will save many otherwise fated patients.

As the incipient stages of cholera are always ushered in by diarrhœa, it stands everyone in hand to watch the condition of the bowels closely and to check any undue laxity without delay.

With an ample supply of *Germ Killer* and the *C. D. Co's Cholera Cure* available for instant use, the housewife can have a much lessened fear of the inroads of the grim monster.

CHILBLAINS.

This exceedingly irritating affection of the feet and hands is caused either by frostbites or too sudden warming of the members after exposure to intense cold. It is more frequent in young children, although many persons are subject to its regular recurrence every winter.

Defective circulation predisposes to these attacks and a course of treatment tending to stimulate the general circulation will usually prevent them. A generous diet, plenty of active exercise and frequent rubbing of the extremities will be of material assistance.

There are several remedies given to relieve the complaint when acute, each of which seems to be efficacious in some cases and not in others. Some one of the list, however, should afford respite from the intolerable irritation in almost every instance.

Should the swelling be unbroken, bathe the feet every night or morning in very cold water or even snow, rubbing them briskly until warmth is restored; or bathe in a strong solution of alum and water applied hot; or apply camphor ice to the parts and promote circulation by rubbing.

Should the chilblains break and ulcerate, keep the sores thoroughly clean by bathing in a solution of warm water and the *C. D. Co's Antiseptine* and dress with folds of linen saturated with the *Antiseptine*.

Persons subject to chilblains should avoid undue exposure and always keep the extremities warmly clothed.

CHAPPED SKIN.

This roughness and soreness of the cuticle of the face or hands and especially the lips is almost universal during the winter season, especially among women and children.

As is generally known it arises from exposure, particularly when the skin has not been thoroughly dried after bathing or other contact with water.

A most pleasant, safe and certain remedy is the *C. D. Co's Rose Cream.* It is a preparation of high medicinal character and at the same time a dainty accessory of the toilet. Applied as directed, it will rapidly heal the severest cases, while its regular use during the colder months will act as a preventive, keeping the skin soft and fair.

It is of the greatest value as an emollient lotion for the complexion and should be used in preference to any other article for this purpoee. While effective it is perfectly harmless and this is more than can be said for most of the preparations for that purpose now on the market.

CHICKEN-POX.

An eruptive disease peculiar to childhood alone, and communicated by contagion almost exclusively.

Children who have had this complaint are totally exempt from further attack.

While during its progress, the patient suffers an intense and disagreeable itching of the eruptions, there is very little, if any, possibility of serious or fatal results. Scratching and abrading the eruptions should be prevented, so far as can be done, and undue exposure guarded against.

Unless other diseases set in to complicate the case, the course of chicken-pox is brief and in a few days after the eruptions appear, the child will have completely recovered.

CANKER SORE MOUTH.

This complaint, which is very common in children, is characterized by the membrane of the mouth and tongue becoming red and inflamed and breaking out in small white ulcers; often the tongue becomes swollen and swallowing difficult. It is exceedingly painful and irritating, particularly to young children.

Remedies.

First and foremost, it indicates an impure condition of the blood and complete cure can be affected only by striking at the root of the evil and correcting the constitutional impairment.

Give the *C. D. Co's Sarsaparilla Resolvent* regularly and the system will soon throw off the foul matter which indicates its presence in these ulcers. Not only in this, but the whole general health of the patient will rapidly increase in vigor.

As a gargle or wash for the mouth, a strong decoction of the common herb known as golden seal is highly recommended. Also strong sage tea has proven efficacious with children.

Adults will find that touching the ulcers frequently

with common black pepper will cause them to disappear rapidly.

CONSTIPATION.

This, as a rule should be considered as a symptom of disease, rather than a distinctive complaint.

Nothing is more injurious than a condition in which the refuse matter, which should daily be thrown off by the system, is allowed to remain and poison the entire physical economy. Let the natural outlets for waste matter became clogged, and serious ailments result.

As we have stated, constipation is an indication only, and nine times out of ten it is an infallible symptom of bilious disturbance and, as such, it is easily remedied.

Stir the liver and digestive organs up to their work, invigorate and stimulate the secretions which are vitally necessary to the smooth running of the human machinery, and the whole system will respond instantly, rebounding from the depression caused by the temporary stoppage.

It may be taken as a safe rule, that when the bowels are moving regularly and smoothly, the general condition of health is all that can be desired. If they are not doing this, if they work too rapidly or not at all, measures should immediately be taken to reestablish the proper tone.

So called chronic constipation is merely the indication of a constant torpidity of the liver and other organs of digestion, and as such is as readily corrected as a temporary or acute case, saving that it may take more time and perseverance.

Remedy.

When constipation arises from biliousness (and it seldom comes from other causes) the intelligent use of the *C. D. Co's Liver Pills* is a positive and certain cure.

They possess none of the disagreeable features of big unwieldy pills, and their action, while firm, is gentle and unaccompanied by the griping and purging horrors of the old style cathartic.

A dose of three or four, either before or after a meal, or on retiring, will rout the most acute indigestion or bilious attack, while if taken regularly, one each night, the most obstinate cases of chronic constipation will surely and quickly yield to their mild but inflexible correction. But a few days of this steady treatment will be necessary before the bowels will move freely and regularly without medicine.

Only those who have suffered the horrors of chronic biliousness, can appreciate the blessings of a perfectly regulated system. With the aid of these invaluable little pellets, there is no one who cannot have a perfectly balanced digestion. Never allow them to be absent from the family medical supplies.

CATARRH.

Aside from the fact that 'this affection of the mucous membranes of the throat or nasal passages may extend to the bronchial tubes and lungs, and thus create an opening for the germs of consumption, in those who are predisposed to that malady, simple catarrh is never attended with fatal results.

Acute catarrh is practically synonymous with a

large proportion of common colds and its symptoms are too well known to need description. Everyone has had a "cold in the head" and can recognize it without instruction.

However it is this very cold in the head which, if neglected, developes into chronic catarrh, one of the most wide-spread, unpleasant, and in every way irritating diseases known, and one of, if not the most difficult to dislodge.

Symptoms.

The intense soreness of the nasal ducts, headache and generally "stuffed up," miserable sensation of acute catarrh, will, if not promptly remedied, run into the, at first less marked but eventually worse, characteristics of the chronic stage.

There seems to be none of the less dangerous ailments which inflicts upon its victims so numerous and varied an assortment of unpleasant, not to say painful, features as this. The constant discharges of foul mucus which invariably accompany it, would be alone sufficient to depress the most sanguine and when, as is nearly always the case, the patient is also subject to almost constant headache, mal-odorous breath, bad taste in the mouth, loss of appetite, and the thousand and one other ills, there is little wonder that it is prominently classed among the wasting diseases.

In its severer forms the patient loses flesh rapidly, and unless the progress of the malady is checked, it is, as we have stated, very liable to superinduce consumption.

Remedies.

It is often stated that chronic catarrh is incurable, but this is a mistake. That it yields with reluctance to ordiary treatment we admit, but the proper combination of curative agents will eradicate it completely.

In all the long list of most excellent prescriptions that the enterprise of the *Consumers Drug Co.* has gathered from various sources and which are offered to their friends and patrons, none is more wonderfully efficacious than their *Catarrh Cure*. The happy discovery of a great specialist, who was finally induced after much effort to give up his secret, for the benefit of the people at large, it has afforded first relief and then complete cure in numberless cases.

While this has a soothing and healing action directly on the diseased membranes, the fact that a long standing case of catarrh causes a terrible drain on the entire vital force of the patient, should not be lost sight of. It is self-evident that it is difficult, if not impossible, for the affected tissues to heal unless supported in their efforts to throw off the disease, by a vigorous general vitality. It therefore follows logically, that the system must be strengthened before the eradication of the disease can be expected.

In almost all cases, there is no better means of doing this than a generous diet, together with a course of the *C. D. Co's* unexcelled *Sarsaparilla Resolvent*. The *Sarsaparilla* not only purifies the blood, which is certain to have become tainted with the poison of catarrh, but it exercises a strong tonic effect on the entire system, especially the stomach and organs of digestion.

In a few instances, where the trouble is exceptionally severe and of very long standing, the whole body becomes impregnated with the disease, and there is in consequence, great emaciation with all the forces at a very low ebb. In this stage, the stomach is usually in a very serious condition and either refuses food altogether, or does not perfectly digest ordinary diet. In these cases the patient should be given the *C. D. Co's Emulsion of Cod Liver Oil*, alternately with the *Sarsaparilla Resolvent*, the *Emulsion* acting as the most effective and highly concentrated tissue-building food that can be had, and at the same time the *Sarsaparilla* tones up and stimulates the organs, as well as drives the impurities from the circulation.

But unless the case be unusually severe, the *Catarrh Cure* and the *Sarsaparilla* will be sufficient. The *Catarrh Cure* should always be at hand as nothing has so quick an effect in disposing of ordinary cold in the head.

Acute catarrh can be cured in a few hours that if allowed to become chronic may take weeks of steady treatment to subdue. A word of warning that the wise will heed.

CROUP.

There are probably few mothers who have been so fortunate as to rear their children to the age of ten years, without more or less experience with this very dangerous disease. And there are thousands of homes that have been forever saddened by the invasions of this dreadful malady. With a few possible exceptions, there is little doubt that croup carries off more young children every year, than any other one disease, to which infancy is subject.

Its attack is so sudden and unexpected and unless relieved, death comes so quickly, that it behooves all mothers to be not only constantly on the alert, but fully armed against it.

Symptoms.

There seem to be few premonitory symptoms of croup that can be relied upon to give warning of its coming. A child may appear in every way thoroughly well on retiring and in two hours awaken suffering from a severe attack.

The almost unfailing characteristic, the harsh, dry, clanging cough, which when once heard is never forgotten, is however a symptom which can be depended upon and a note of warning which should awaken the mother to instant action.

The patient may cough several times before awakening, but in a very brief period the child will start up gasping and struggling for breath. This sense of suffocation accompanied by hissing, wheezing respiration, together with great distress, a very hot skin and rapid pulse are invariable symytoms.

Remedies.

When possible, a skilled physician should be sent for at once, for while the attack can usually be controlled by the prompt action of the mother, yet this cannot always be depended upon and in very obstinate cases the services of a surgeon may be required. But the efforts of the mother should not be relaxed or neglected for a instant, even if the doctor be within easy reach, for children have frequently died within a very few moments after the first symptoms became evident.

The patient must be removed to a warm room and wrapped up so that there is not the slightest chance of draught or exposure. A plentiful supply of hot water must be had immediately and arrangements be perfected for administering a hot bath if necessary. First, however, the patient should be given a dose of the *C. D. Co's Ipecana*. This admirable compound which contains, with other very beneficial agents, a large proportion of the wine of ipecacuanha, acts as a gentle emetic and should induce vomiting in a short time. If the first dose is not effective, repeat at intervals of about fifteen minutes.

While vomiting is one of the quickest and most certain means of relief, all harsh emetics should be strictly avoided. The *Ipecana* is a remedy that cannot be surpassed because of its mild, though positive action and also from the reason of its exerting a strong relaxing and sweating influence, both effects being highly beneficial. Moreover, it is absolutely harmless to the weakest infants when given as directed. All "hive syrups" and ordinary compounds of ipecac should be shunned because of the crudity of their mixture and the proportion of antimony which most of them contain. *Ipecana* is a very delicate compound of efficacious agents highly refined and adapted to the fragile systems of very young children.

It is not always necessary that vomiting be produced, as the relaxing effects of the *Ipecana* will frequently induce soft easy breathing and consequently relief, without the stomach having responded.

As accessory measures, the child may be allowed to inhale the steam from hot water, be immersed in a hot bath, or the throat wrapped in cloths wrung as dry as possible from very hot water and constantly renewed.

This treatment will prove successful in nine cases out of ten and should it not do so, the physician will have to adopt measures, which can be used with safety only in professional hands.

Mothers should always remember, that after an attack of croup, the child is very liable to its recurrence at every slight cold, therefore great care should be exercised both in permitting exposure in bad weather and seeing that the child is warmly clad. This caution will prove well rewarded.

DIPHTHERIA.

In its malignant form this disease has inspired such general dread that extended mention of its deadly character seems unnecessary. Fortunately, however, the simple cases which are comparatively easy of treatment far outnumber those of the more contagious and fatal form, save in instances where it is epidemic.

It would be as criminal for the author to advise household treatment of this very serious ailment, as it would be useless in the reader to attempt to combat it, unaided by an expert physician; therefore he will confine himself simply to pointing out its symptoms and causes and suggesting preventive measures.

Symptoms.

The characteristics of diphtheria most easily discerned by unprofessional eyes are first, general feeling of lassitude and indisposition, sometimes followed by slight nausea before the disease has become apparent in the throat. Its presence at this center of local attack is usually the first symptom which can be recognized with any degree of accuracy, by one unskilled in diagnosis.

The throat becomes rapidly inflamed and swollen and a little later the membranous lining is more or less thickly covered with the grey or white patches, which are the peculiar feature of the disease.

The patient should have been under a physician's care before this stage is reached, but if not, one should be summoned without delay. Diphtheria will often test the knowledge and skill of the best medical men and no housewife should attempt to treat it.

Preventive Measures.

Diphtheria, particularly of the malignant sort, is another of the appalling list of diseases which have been proven to be communicated through germs or microbes. Long as this catalogue already is, the microscopists are constantly making additions to it and thus clearing up the mysteries which have long surrounded the origin of so many ailments. These great discoveries not only satisfy the long baffled curiosity of the profession, but they have proven of untold value to the world at large, through enabling science to combat and destroy the poisonous atoms, before they can create disease in the human system.

The resources of chemistry have been invoked for agents, penetrating and powerful enough to annihilate these insidious, though invisible, messengers of death and as the result of months and years of experiment and research, the chemist has supplied that need. None of these disinfecting compounds is excelled by that famous preparation of the *C. D. Co.*, the *Germ Killer*. It is quick and certain death to the whole poisonous brood of microbes and as such, is an invaluable agent in the prevention of disease.

In common with the entire group of bacilli, the diphtheria germ multiplies and flourishes in waste and decaying matter and therefore it is extremely important in diphtheria epidemics, that all receptacles for refuse should be kept thoroughly clean and disinfected.

Not only in epidemics of all diseases, but *at all times* should this vital precaution be observed. Never forget that the old saying "an ounce of prevention is worth a pound of cure" is one of the proverbs which has proven its wisdom in millions of instances. Keep your surroundings in proper sanitary condition and danger in any epidemic will be reduced to a mere possibility.

DYSPEPSIA—INDIGESTION.

This curse of the American people seems to be a peculiarly national complaint.

The cause for this is readily apparent to any thinking observer. In no other civilized country is there the universal recklessness in breaking every known law of health as in this, and none of these rules are so generally trampled upon as those relating to the stomach.

Almost without exception every culinary compound, that is so widely popular as to be known as a distinctively American dish, is a mixture of the most indigestible articles of food imaginable. Add to this the fact that the average American, especially if a business man, is living under a terrific pressure, driving every energy to its utmost, without the least regard to the proper and necessary periods for rest and relaxation, and it is wonderful that we have any percentage whatever of sound, healthy men, women or children.

Popular opinion to the contrary notwithstanding, the stomach and its auxiliaries of the digestive apparatus, is the head and center of the human system and from it arises nine-tenths of the ailments to which mankind is heir.

Therefore, it stands everyone in hand to keep this important factor of his health and happiness in proper condition. And this can be accomplished only by careful judgment and moderation, both in the selection of food and regulation of the habits.

It is not our purpose to enter into a lengthy discussion of the various articles which go to make up the food and drink of mankind, pointing out those that are most digestible and nutritious or the opposite. No general rules can be laid down on this point that would be worthy of consideration. What one person finds highly palatable and sustaining would cause misery to his neighbor and *vice versa*. Every individual who has reached years of discretion and is of sound mind is capable of selecting those articles of diet which experience has proven best adapted to his personal digestive peculiarities, or of refusing those which exercise a deleterious effect.

If this intelligence be applied, no one need have fear of that many-armed, ever threatening octopus, dyspepsia, but if the stomach be wilfully abused by forcing it to attempt to digest those food products at which it has heretofore often rebelled, it is only a question of a short time when it will become so weakened as to perform its work but indifferently, under the most favorable circumstances.

The best method of curing dyspepsia is not to have it. Unless, as in rare instances, the sins of the father be visited upon his children, and a weak stomach is inherited, caution and a certain amount of abstemiousness are certain preventives. Remember that the laws of nature can be broken with apparent impunity for a time, but eventually retribution will surely come.

A word to those unfortunates who are already suffering the torments that come only from weakened or diseased organs of digestion.

The simple case of over-eating, while unpleasant at the time, will usually disappear in a brief period without the use of medicines. But well developed chronic dyspepsia is as difficult to overcome as the simple instances of acute indigestion are easily relieved, and those sufferers who would free themselves from this affliction must be prepared for a long continued, unflinching struggle, no matter what discouragements may at first confront them. The result of years of indiscretion cannot be undone in a day, although with the aids which science has to offer, it will not take many weeks.

Remedies.

The ordinary case of acute indigestion, arising from

forcing the stomach to hold more than it is prepared to digest, or from eating rich and very indigestible food, can be quickly relieved by taking one of the *C. D. Co's Liver Pills* immediately after eating. These pleasant little pellets are designed not only to act on the secretions of the liver, etc., but to directly stimulate the stomach, in fact to have largely the effect of the various preparations of pepsin which are now so popular. They have the inestimable advantage, however, of being composed of purely vegetable agents and are absolutely harmless. In doses of one pellet taken immediately after eating, they do not have a cathartic action, but simply aid the digestive organs in disposing of their unwonted load.

Chronic dyspepsia, however, requires different treatment and first and most important of the remedies which the dyspeptic who would be cured must adopt, is a very careful adjustment of his diet to the actual need of the system.

Eat only those articles of food which have proven themselves in the highest degree acceptable to the stomach and which also contain a large percentage of nutriment. Carefully avoid excesses of every description, eating at regular periods and drinking little or no malt or spirituous liquors.

Overwork and lack of the proper amount of sleep are frequent causes of digestive weakness. What will give rise to any ailment will certainly aggravate it, therefore, no relief of chronic dyspepsia is possible, unless the patient regulates his habits and affairs to conform with the recuperative necessities of his system.

The mere observance of these regulations will not suffice, unless the digestive apparatus is afforded the assistance of a tonic stimulant, and it is in this that the allied sciences, medicine and chemistry, proffer their aid.

Dyspepsia is simply a weakened and diseased condition of the stomach and this weakness and disease must each be given the treatment necessary to remove them.

Favoring the stomach in every way possible by the selection of easily digested, nutritious food, enables it to recover its lost strength, providing it is toned up and the diseased tissues healed by the action of the *C. D. Co's Sarsaparilla Resolvent.*

Sarsaparilla Resolvent is not alone a blood purifier, but a powerful general tonic, whose benign action is felt throughout the entire system. Naturally the stomach receives this effect first and very markedly and consequently responds without delay.

This treatment will eventually completely cure the most confirmed and long-standing cases of dyspepsia, and we earnestly solicit a trial of it, especially from those sufferers who have met with frequent disappointments in using other remedies. We know they will receive the most happy results from this.

In cases where there are pronounced symptoms of chronic biliousness as well as weakness of the stomach, the regular use of the *C. D. Co's Liver Pills*, taken one or two each night on retiring, in connection with the *Sarsaparilla Resolvent* at meals, will be found extremely beneficial. This will not be necessary for more than a very few days, after which the *Sarsaparilla Resolvent* will be all the medicine required.

The *Liver Pills* are recommended solely to stir up the old, long-settled accumulations of disease and thus expedite recovery. The *Sarsaparilla* will achieve the same results unaided, but as its action is somewhat different it will take longer, hence the *Liver Pills* add materially to the quickness with which a complete cure can be arrived at.

DIARRHOEA.

This excessive looseness of the bowels is very often only a symptom of organic derangement, rather than a disease *sui generis* although the form known as summer complaint, as well as diarrhœa arising from irregularities in diet, are the simple results of local causes and can be treated accordingly.

Moreover, diarrhœa should not be confounded with dysentery, a disease which it resembles in many respects, but which requires radically different action to afford relief. In this connection we will consider merely the complaint in its common distinct phase, the one in which the ordinary remedies available in every household will afford relief. In those instances where diarrhœa is a symptom of a graver disease, the other characteristics will be so marked as to prevent any misunderstanding.

It seems hardly necessary to remind our readers of the importance of checking such a debilitating and weakening ailment as this, or to impress upon their minds the fact that if not relieved, it quickly runs into chronic diarrhœa, one of the most difficult diseases to permanently cure, known to the medical profession. Moreover, death has resulted from neglected cases of simple diarrhœa.

Causes.

The ordinary form of diarrhœa arises most frequently from exposure to sudden extremes of heat and cold, such as heating the body by active exertion, then too quickly cooling it. Hot days and chilly evenings also bring it on very often. No cause is more common, however, than the eating of unripe fruit, and other indigestible articles which irritate and inflame the membranous lining of the bowels.

Symptoms.

The intense griping pains of cholera morbus and colic are a certain prelude of diarrhœa and should receive the prompt attention noted in the chapters on those complaints.

Often simple diarrhœa is preceded only by slight sensations of uneasiness in the abdominal region and a feeling of lassitude accompanied by headache.

Frequently, however, the first symptom is the loose, watery evacuations which are the main characteristic of the complaint.

Remedy.

Should there be no indications of dysentery (for which see next chapter) all forms of diarrhœa should be gently but firmly checked without delay.

Give the *C. D. Co's Cholera Cure* as directed on label, at intervals of about an hour and unless the case is unusually severe, two or three doses will entirely check the discharges. It is also advisable to keep the patient in a reclining position as much as possible, thus facilitating recovery.

DYSENTERY.

Too many people confound this very serious disease with simple diarrhœa, endeavoring to relieve it with the same measures which would apply to an ordinary case of summer complaint.

No error can be more dangerous, since the treatment in these two ailments are almost exactly opposite. Furthermore, dysentery is a disease which demands the instant assistance of a skilled physician and no amateur should endeavor to handle it unaided.

This being the case, the author will confine himself merely to endeavoring to point out the characteristics by which this complaint may be recognized and the features which distinguish it from diarrhœa.

Symptoms.

Dysentery is almost invariably preceded by intense griping pains through the abdominal regions, occasionally accompanied by chills and fever. This is usually quickly followed by the distinctive symptoms of the ailment, the passing of scanty watery discharges from the bowels, usually more or less mixed with blood and mucus. This is always accomplished with great straining and pain to the patient, there being an almost constant desire to evacuate the bowels, while the discharges are very scanty and often times the sufferer cannot pass anything even after great effort. In this, lies the point by which the laity can positively recognize dysentery. A person afflicted with ordinary diarrhœa can without difficulty free the bowels of their contents, the discharges being very complete and almost invariably affording temporary easiness and relief. When this is the case, the remedies suggested in the previous article will cure in a short time, but if the symptoms of dysentery are at all marked, no time should be lost in securing the services of a good physician.

There are of course many indications of this disease in addition to this most pronounced symptom above noted, including the quick pulse and general heat of fever, vomiting, etc., but the enumeration of them would tend to confuse the reader rather than enlighten him. They are but the signs of the general constitutional disturbance which so grave a malady always produces. The *tenasmus* or straining at stool is the danger signal to be heeded.

Pending the arrival of the physician, the housewife can do little besides keep the patient in bed, the recumbent position tending to alleviate the pain, and use such means to give relief as circumstances afford.

The attending medical man will direct the diet and accessory treatment to suit the individual requirements of the case. One final caution, never give an astringent in dysentery as in diarrhœa.

INFLAMMATION of the EYES.

Inflammation or what is commonly known as "sore" eyes, are remarkably frequent with all classes, especially among children.

Except when the inflammation arises from an accident or injury, such as the introduction of hard, foreign substances under the lids, this affection is traceable directly to a scrofulous or other taint or impurity in the blood, omitting, of course, occasional instances in which a temporary inflammation of the tissues surrounding the eyes results from some severe constitutional impairment, as fevers, etc. Almost every ordinary case of this kind, however, is an indication of blood humor and should receive corresponding treatment.

Remedy.

A local application to disperse the inflammation is a necessity and none is better for the purpose than the *C. D. Co's Prescription No. 37.* It is a preparation which has long been used by a noted oculist with very happy results and it is certain to relieve all ordinary cases of this distressing complaint, with almost marvelous quickness. Those who are subject to these disturbances would be wise to keep a supply of this most valuable compound, against their time of necessity.

Local applications are of infinite advantage in affording temporary relief, but, as in every other known disease, to eradicate it, the roots must be destroyed. The healing effect of a good lotion which drives away the inflammation should not blind one to the importance of removing the cause of the trouble.

Chronic cases, or periodic attacks of inflamed eyes are convincing evidence of the necessity of a thorough course of blood purification. The *C. D. Co's Sarsaparilla Resolvent* will accomplish this quickly and completely, no other remedy being required.

In discussing this most excellent compound in its

relation to other disorders requiring alterative treatment, the author believes that he has said sufficient to impress the reader with its pronounced good qualities. But all that he has written, or can write, will not equal one-thousandth part of the vast volume of praise that rises up from the tens of thousands who have been benefited by it. It is a preparation which should be ever present in every household in the land.

ERYSIPELAS.

An exceedingly deep seated and obstinate complaint, characterized by an intense inflammation and redness of the skin in patches or blotches, of greater or less extent, according to the severity of the attack.

Causes.

Many writers seem to be slightly in doubt as to the origin of this disease, although the majority and the very highest authorities, agree in the opinion that it arises from an impure condition of the blood. The immediate cause of an attack of erysipelas may be any of a great number of circumstances or conditions.

Exposure, indigestion, intemperance, injuries or wounds from which come local inflammation that rapidly runs into malignant erysipelas. Any of these or a hundred other matters may be the active means of developing this malady. In fact anything that stirs up the latent poison in the blood will bring it on.

The larger portion of those who suffer from this disease are subject to periodic attacks, as they must ever be until they adopt a course of treatment which will thoroughly renovate and build up their system, and eliminate from their blood every trace of the complaint.

There is not the slightest doubt that erysipelas is transmitted by inheritance, as are nearly or quite all of the humors of the circulatory system. Parents who have shown even ₁slight traces of this disease, should watch their children carefully, or better yet, use the means which science offers so freely to thoroughly drive out the poison.

Symptoms.

Acute attacks of erysipelas are usually ushered in by a feeling of lassitude and general indisposition accompanied by low fever and loss of appetite. Sometimes intense thirst, nausea, and diarrhœa are in evidence.

Usually the characteristic feature of this disease, the inflamed, painful blotches on the skin make their appearance very soon after the premonitory symptoms are noticable. These patches are most commonly found about the head or face, although they may occur on any portion of the body. Frequently they make their first appearance near the nose or ear, or if there be any scratch, cut, or other abrasion of the skin, causing light local inflammation, the incipient blotches are likely to originate at that point.

The exact location of the first appearance, however, is soon lost sight of as the inflammation, swelling and discoration extends rapidly, soon covering a considerable area. With this increase of external symptoms the general derangement grows more pronounced, until the patient is in a very bad condition.

Acute erysipelas is another of the long list of diseases in which the housewife must invoke the aid of a physician, limiting her own treatment to the accessory measures indicated by him. There are few ailments which tax the strength and resources of the nurse more than this, and she will find ample opportunity of caring for the patient, even with the assistance of the doctor. Consequently, there should be little delay in sending for a physician when the evidences of erysipelas are noted.

Remedies.

As has been stated, the treatment of acute erysipilas must be left to the medical attendant, therefore, it would be profitless to discuss it here. Only a word of caution—do not use any local applications until the doctor arrives, save a little dry flour or starch. Avoid all wet and chilling local treatment as far as possible, as the concensus of opinion among specialists in this disease is that their use is more often attended with injury than benefit.

Advice as to the management of acute cases, however, is an unimportant matter in comparison with instruction how to prevent these attacks and *remove* the disease altogether from the system.

Unless this done there is the ever present possibility of its breaking out and it is not necessary to remind the reader that fatal terminations of this malady are far from rare, without regarding the intense suffering which charcterizes its acute form, even if finally for the time being subdued.

No amount of local treatment will have the slightest weight in eradicating the poison and only a perserving unfaltering course of alterative measures will accomplish it. The impaired blood must be purified before immunity can be assured. Nothing will more quickly and surely drive out erysipelas poison than that unexcelled blood renovater the *C. D. Co's Sarsaparilla Resolvent.* Do not wait for the disease to assume its acute form before taking these means of cure. If you have ever suffered from it you will surely be again the victim unless it is forever driven from your blood. Begin taking the *Sarsaparilla* now and a short time will suffice to free you forever from this fiery fiend.

Furthermore, its strong tonic effect will prove very grateful even though you may not feel any prominent symptoms of disease. As a general invigorator it is without an equal. Insure your children also against the outbreak of this latent imp that will, if not slain, develop into a full grown devil to torture them when least expected. If you have been its victim, there are very few chances that your offspring will escape if eradicating measures be not taken.

Finally should you not heed these words of warning and neglect this safeguard until after the complaint has developed its burning torment, always take the *Sarsaparilla Resolvent* after your physician has subdued the attack and if the treatment is conscientiously persevered in, you need have no fear of its recurrence. A word to the wise should be sufficient.

ECZEMA.

Of the very extended catalogue of cutaneous or skin diseases there is not one more frequent, more unpleasant, or harder to completely eradicate than this. All ages and classes are subject to it and in common with all ailments which have their origin in impurities or taints in the blood, it is transmitted directly by inheritance.

It seems to the author unnecessary to devote much space to the description of the symptoms of this disease, as the moist scaly eruptions which cover more or less thickly various portions, or often the entire surface, of the body, are easily recognized by everyone.

Remedies.

This, however, is a more important matter and one which deserves extended mention.

There are scores of so-called specifics for all skin diseases consisting of lotions, ointments, etc., widely advertised and correspondingly used by the victims of this and its kindred ailments. Such of these as have any merit whatever often allay the irritation to a certain extent and for a temporary period, but an external application can no more remove a blood humor, such as this, from the system, than it can cure consumption or dyspepsia.

A soothing emollient such as the *C. D. Co's Rose Cream*, is often of value in eczema as an accessory measure only, to quiet the irritation and soften the hard scales of the eruptions, but the disease itself can be eradicated only by driving it from the circulatory system with a powerful purifying alterative.

This is one other of the manifold instances in which the *C. D. Co's Sarsaparilla Resolvent* has proven its infinite value as a remedial agent. In all cases where the vitality and purity of the blood is in any way impaired, it is an absolute specific, neither excelled or equalled by any preparation on the market.

We recommend it as an unfailing remedy for eczema and other skin diseases and a brief trial will fully substantiate the truth of this assertion. If you are a sufferer from this loathsome disease do not refuse this certain means of quick return to health and happiness, no matter how many and disheartening may have been your previous disappointments. This is one preparation which is put up with the honest intention of doing good and no expense has been spared to achieve this end. Were this not the case, the manufacturers would not trust to the merits of the article itself to make its own name, rather than create a false reputation through skillful advertising.

FELON.

An inflammatory affection of the extremities, usually occuring on the fingers, thumb, or hand.

This exceedingly disagreable complaint is ordinarily first indicated by a severe, deep seated, prickly, throbbing pain which rapidly increases as the swelling advances, until it becomes at times almost unbearable. The bunch or swelling will finally break and suppurate profusely unless lanced.

It is always best to invoke the aid of a surgeon as

soon as the felon is sufficiently advanced to make its identity certain, for if these pests are neglected there is a possibility of their causing blood poisoning with serious results.

Lancing them early in their development will often check their course, but this is an operation which should never be attempted by an amateur, a physician alone being capable of performing it properly.

After the felon is lanced or has broken naturally, it should be dressed with the *C. D. Co's Antiseptine* only. This preparation will heal the sore quickly, prevent the formation of proud flesh, allay inflammation, in fact it is an ideal dressing for all wounds and sores.

FEVER SORES.

No dressing will afford such satisfactory results for fever sores as the *C. D. Co's Antisepine*. All ulcerous eruptions require an antiseptic healing lotion and the *Antiseptine* will be found on trial to be the most effective preparation of this class obtainable.

GRAVEL.

Stones in the bladder or kidney is a malady which tests the skill and knowledge of the most experienced physician and no pains or expense should be spared in securing the very best professional aid, when this disease is suspected.

Causes.

Calculous disease of the kidneys is at least developed, if not originally caused, by over indulgence and excesses both in eating and drinking, sluggish habits and lack of proper exercise.

Symptoms.

While there is no disease in which there is so wide a variety of symptoms in different cases, several, if not all of the following indications may be recognized in the majority of instances.

Feeling of indefinable discomfort and apprehension accompanied with slight pain through the region of the kidneys.

Shooting pains more or less severe, arising in the same portion of the body and radiating to other parts often to the groin, thighs or hips.

A deep seated, steady pain in one or both of the hips.

Frequent desire to void the urine, accompanied with much pain and often containing small particles of stone or a lime like substance.

These are of course only warnings of the intense agony which invariably attaches to the passage of a stone of any size. The terrible anguish of these attacks can be appreciated only by those who have suffered from them, as it is beyond description.

Only too frequently they terminate fatally and those who have the least premonitory symptoms, should lose no time in placing themselves under the care of the best physician at hand. No directions for self treatment will afford relief or avert disaster.

GOUT.

This quasi-rheumatic affection of the lesser joints, is far less common in this country than in many others, particularly in Great Britain. The predisposing cause of the complaint is of itself sufficient explanation of this phenomona.

Barring the fact that a strong tendency to this disease is unquestionably transmitted from one generation to another, all authorities unite in the statement that gout is caused by excessive eating of rich foods together with the constant use of liquors, especially wines and malt drinks. Those whose mode of life is sedentary or who are of a sluggish temperament are much more liable to the complaint, other circumstances being equal.

While there are thousands of cases of gout in the United States, the average American is too lively and energetic to succumb, no matter how much his eating and drinking may tend to the development of the complaint.

Like all rheumatic maladies, gout is primarily a disease of the blood, arising from what is technically known as uric acid in the circulatory system which is deposited around the joint locally affected. This is almost invariably the ball of the great toe, although occasionally the entire foot is affected.

Remedies.

While the disease may seem to be entirely seated in the extremities, it does not by any means follow that it can be cured by local treatment, for in fact all applications, etc., are not of the slightest benefit in eradicating the disease.

As has been stated, gout and all other complaints of a rheumatic nature are located in the blood and an alterative treatment only will remove them.

A good liniment as the *C. D. Co's Anodyne Liniment* applied to the inflammation will often afford relief, but it will take a powerful blood remedy to effect a permanent cure.

After all that has been said regarding it, we feel that the reader should turn instinctively to the *C. D. Co's Sarsaparilla Resolvent* when there is need for an agent of this character. In every case where an ailment is traceable to impairment or a diseased condition of the blood, the *Sarsaparilla* will afford quicker and more marked results than any pharmaceutical preparation available. Try it and be convinced.

HEADACHE.

Headaches of every character are symptoms only, not local ailments. The general disturbances which they may possibly indicate, include almost the entire list of the graver diseases to which the flesh is liable. They are among the premonitory indications of fevers, dysentery, catarrh and nearly all serious complaints.

Manifestly it is unnecessary to consider this class of headaches at length and the author proposes to dwell only briefly on the common forms known as sick and nervous headache.

As has been noted in the author's remarks on biliousness, ninety per cent. of ordinary headaches have their origin in temporary derangement of the digestive organs and, this being the case, there is not the slightest excuse for their continuance.

The *C. D. Co's Liver Pills* are a specific within the reach of everyone and there is not a single case of occasional or chronic sick headache which will not be compelled to yield to their action, if they are used conscientiously as directed. Those subject to sick headaches are referred to the chapters on biliousness and indigestion, and if they will follow the advice there given, they can be assured of immunity.

Another frequent form of headache is that arising from a weakness or degeneration of the nervous system. This affection has may peculiar features indescribable in a written article, but which are only too easily recognized by those affected with so grievous a disease as this. Naturally the only remedy is to invigorate and build up the nervous system and until this is done, the patient can expect only more and more frequent attacks of excruciating pain.

For all those whose nerves have been weakened or diseased through overwork, or any cause whatever, the *C. D. Co's Nervita* is a boon which should be accepted without question or delay. While a powerful nerve tonic, it is at the same time a builder of nerve matter, its action being to replace the waste substance as well as to eradicate the disease. Its reviving, stimulating effects on a worn-out system are little short of miraculous. As a local accessory measure the use of the *C. D. Co's Aromatic Ammonia* will be found most grateful in cases of headache. Bathe the patient's temples and forehead and allow her to inhale the fragrant salts and entire relief will frequently follow.

Do not, however, forget that headaches indicate a deeper disturbance, which must be corrected to insure immunity from further attacks.

HIVES.

This very irritating complaint which is more common with young children and girls than with adults or males, is very rarely dangerous and can be entirely relieved with little difficulty.

Causes.

Except in a very few instances hives or nettle-rash is caused by a form of indigestion either acute and temporary, or in a measure chronic. Naturally if long continued, it involves more or less derangement of the blood, but in any case its cure, is a matter of a few days only.

Symptoms.

The characteristics of this disease, the broad flat eruptions or *wheales* rather pale in the center, deepening in color to the edges and accompanied with almost intolerable burning and itching, are easily recognized. They occur more frequently about the shoulders, back or loins and invariably increase both in size, number and painful sensations toward evening and when the patient first retires.

Remedies.

Given the fact that this complaint arises from defective digestion and its remedy is self evident.

Care should be exercised in the selection of food and those articles which have proven themselves not grateful to the stomach should be carefully avoided.

Then the usual medicinal agents for the relief of indigestion, the *C. D. Co's Liver Pills* in laxative doses of one or two on retiring, will bring relief. If the case be chronic, the attack appearing frequently, the *Sarsaparilla Resolvent* should be taken regularly. In fact the *Sarsaparilla* will prove of great benefit in acute cases as well and if taken regularly for a few days will prevent further trouble of this nature.

To allay the intense itching of the eruptions, they should be kept anointed with the *C. D. Co's Rose Cream.* Common or table salt rubbed on, also the application of flour or starch will sometimes ease the irritation.

HAY FEVER.

This peculiar form of catarrh, asthma or a combination of both, has long been a problem to the medical profession on both sides of the ocean.

Numberless theories as to its origin and the cause of its peculiar periodic appearance have been advanced, some of which seem based on sound reasoning and others are on their face the wildest vagaries of their author's imagination. None of these opinions have been proven to be in every case accurate and the writer under these circumstances prefers to confine himself to effects and remedies, rather than be committed to one of the many theories of the cause of this disease.

Symptoms.

In many ways hay fever resembles common catarrh of the head and nasal passages, being characterized by sensations of heat, fullness and generally unpleasant condition of the head, with itching of the eyes, nose and throat, sneezing, etc., and with it always comes a feeling of lassitude and irritability.

Often the asthmatic symptoms of compression of the chest, spasmodic difficulty in breathing, etc., are quite marked.

True hay fever, however, differs from both these maladies in its distinctive periodic visitations. It is known during the hot summer months only, and almost invariably attacks the patient regularly every season, often on the same date each year, disappearing as suddenly as it came, about the same time every autumn. It is this singular feature of the complaint which has mystified the medical fraternity and made it so difficult to exactly determine its pathology.

Fatal or even serious terminations of hay fever are very rare but the patient is often entirely unfitted to perform his daily duties and in severe cases life is almost a burden to the sufferer.

Remedies.

Very naturally in a disease whose nature is so hard to analyze, a specific which can be relied upon, is practically among the things impossible.

Change of residence, especially to some sea side or lake resort, seems to be the most successful measure so far discovered, to alleviate the discomforts of hay fever. It is doubtless aggravated by the dust common to inland localities, during the summer, since those patients who have been able to spend the season near or on great bodies of water have either been entirely free from the disease, or the severity of its attacks have been greatly mitigated. Unfortunately, the great body of sufferers from this unpleasant ailment are unable to avail themselves of this means of relief, and to these a few suggestions which may be of benefit are given.

Several of the *C. D. Co's* standard preparations have been employed with great benefit in numerous cases of hay fever, but the author does not by any means recommend them as infallible, for no medicine compounded is a certain relief. If the victim of this complaint feels willing to try them on the possibility of their benefitting him, the chances are largely in favor of his being more than pleased with the result but there is also room for disappointment, should his case be one of those which baffles the most penetrating compounds of the *materia medica*.

The patient who wishes to make this effort should begin several weeks before the date at which the yearly attack is expected, by taking the *Sarsaparilla Resolvent* regularly. Since this ailment combines the characteristics of catarrh and asthma, both of which diseases imply a more or less depleted condition of the blood, this preliminary course of tonic and invigorating treatment, will very often enable the system to resist the usual attack of hay fever altogether, or will almost invariably lessen its severity. The experiment is well worth trying, while past results have demonstrated that it is more of an assured success than a venture, in the great majority of instances.

The *Sarsaparilla Resolvent* should not be dropped

even if the disease manifests itself, for it is bound to have a great effect in mitigating the intensity of the disagreeable characteristics, if it should not remove them entirely.

To allay the inflammation of the membranes of the nose and throat the *C. D Co's Catarrh Cure* will be found of infinite benefit. Its soothing, healing effects will prove most grateful to the patient, quieting the irritation and clearing the head of that stuffed, aching sensation so intensely unpleasant and exasperating.

Should the asthmatic symptoms be pronounced, causing the patient great difficulty in breathing, especially during the night, the *C. D. Co's Asthma Pastilles* will usually remove them, affording restful uninterrupted sleep.

If you have tried many other remedies for hay fever without satisfactory results we ask that you give these a trial also, for while, as we have stated, they are not guaranteed a specific, yet they have helped hundreds of cases and there is every reason to believe they will help you.

HOARSENESS.

The form of hoarseness which attends a cold or acute bronchitis will disappear with that complaint and the *C. D. Co's Cough Balsam* is a certain cure for both. The hoarseness to which singers and public speakers are subject is another matter and should receive entirely different treatment. The *C. D. Co.* compounds a specific for this annoyance called *prescription No. 64*, which if clergymen, singers, etc., would once try, they would never be without. It is put up in the form of tiny tablets, a small bottle of which can be easily carried in the vest pocket and they are guaranteed to keep the voice clear and strong, as well as relieve all pain and soreness of the vocal organs almost instantly.

A card to their nearest representative or to the laboratory brings them.

HEARTBURN.

This popular name for an unpleasant sensation is a great misnomer, since it arises from the stomach alone and has no connection with heart. It is a symptom of indigestion merely and can be temporarily relieved by taking either powdered charcoal or prepared chalk, which can be procured at any drug store.

But it is evidence of a disordered stomach and the only wise treatment is to tone up and invigorate the digestive organs and consequently the entire system, as indicated in the chapter on indigestion and dyspepsia.

It is invariably caused by some article at which the stomach rebels and by using a little judgement in avoiding those substances that experience has shown as creating this acidity of the stomach, attacks may be entirely averted.

DISEASES of the HEART.

All affections of the heart are of such a delicate and

serious nature that they call for attention of a skilled physician and any advice as to home treatment would be highly out of place in this work. They would tend to confusion and possibly lead to delay in seeking the services of a doctor, which might prove disastrous. The symptoms are very often indistinct and in such great variety that even experienced specialists are puzzled. Therefore, heart diseases are very far removed from the sphere of the household healer.

Palpitation, however, may result from indigestion, over-exertion, excitement or a debilitated condition of the entire system. It is frequently found in cases where the heart itself is entirely free from any organic disease and is, in such instances, a symptom of other disturbance.

When caused by indigestion or dyspepsia, the correction of the superinducing ailment will entirely remove the apparent affection of the heart.

But the fact that severe exertion or excitement brings on an attack of palpitation, is convincing evidence that the general strength and vitality of the system is impaired and this warning should be heeded, or more serious results may follow.

To supply this waste of vigor and build up the tissues, the *C. D. Co's Emulsion of Cod Liver Oil* is unexcelled and its beneficent action is almost instantly apparent. If preseveringly used it will usually entirely remove all signs of palpitation.

Nature always gives ample warning if her action or forces have in any way deteriorated. These danger signals are many and varied, but the knowledge to recognize them is easily acquired and for this purpose only, has the author labored in preparing this little book.

But if after his readers have been taught to know these indications they do not heed them, then his work in writing and their time in perusing have been utterly wasted. Every word of advice as to what to do and what to use in this perpetual struggle against disease and death which is constantly being waged by the entire human race, has been offered in perfect good faith and with the earnest conviction that if the suggestions be followed the reader would be greatly benefited thereby.

If those who are themselves ailing or have others in their family or neighborhood who may be suffering in the coils of disease, refuse to believe and follow these directions, which are the result both of long experience and extended research, they will themselves suffer more deeply than the writer.

On the other hand if in one instance his work has proven the means of alleviating ever so little the pain of one sufferer only, among the untold millions in the world, he will feel amply repaid for whatever labor and inconvenience the compiling of this little book may have cost.

HOOPING COUGH.

This is one of the most distinctive of what are known as children's diseases and there is probably no mother who has not been called upon to witness it. There are very few, if any, children who escape having it and fortunately it very rarely attacks the same person the second time.

It is very seldom that simple hooping cough uncomplicated with other diseases results fatally, but the spasms of coughing which are its chief characteristic, cause the little one infinite suffering and everything possible should be done to mitigate them.

The exact origin of hooping cough has never been clearly determined, although it is unquestionably contagious to a certain extent. All children seem to have the one struggle with it and it is doubtful if any preventive measures would succeed in warding it off altogether.

Its symptoms are so well known to every one that we will not refer to them, but consider the available

Remedies.

The writer questions whether there is any preparation which will positively cure hooping cough, being of the opinion that the disease is bound to run its course in a measure.

But there are many agents which will greatly mitigate the severity of the ailment and relieve the terrible spasms of coughing. None of these will prove more satisfactory than the *C. D. Co's Cough Balsam*. It should be given carefully as directed on the label, as it is a strong compound and the doses for young children must be proportioned to their age and strength.

This will be the only medicine advisable, since all that can be done is to relieve the severity of the paroxysms.

The child should be watched carefully and undue exposure be avoided, although when properly clothed, a little exercise in the open air on all but very stormy days, will be rather beneficial than otherwise. The patient should also be shown a great degree of consideration, as excitement and anger will tend to bring on and greatly aggravate the spasms of coughing.

While the disease may not be very serious in itself, it should not be held and treated lightly, as complications are frequent, which add much to the gravity of the ailment. These can usually be prevented if proper care in the management of the simple case of hooping cough be shown.

ITCH.

This very loathsome skin disease does not appear to be nearly so prevalent in the older settled portions of the country as formerly, or as it now is in those localities where society is yet in its primitive state.

Occasional cases are, however, not uncommon and the highly contagious character of the ailment makes it most important that it be quickly and thoroughly stamped out.

It is communicated only by contact with the person or clothing of the patient and it has long since been proven to be caused by the ravages of a minute insect which imbeds itself in the skin, causing the well known scabs or eruptions which give the disease its scientific name, scabies. The frightful itching which attend them explains the common term by which it is designated.

Remedies.

While this disease very rarely impairs the constitution to any serious extent, yet it cannot but have more or less effect on the general health and especially the blood.

Consequently *Sarsaparilla Resolvent* will be found of great benefit and should always be given in connection with external applications to destroy the parasites which are the primal cause of the difficulty.

Of these ointments one containing a good proportion of sulphur has proven most effective and the following formula which may be filled at any drug store, will be found to rapidly heal the eruptions and eradidate the disease, providing the internal treatment with *Sarsaparilla Resolvent* be pursued:

Benzoated lard, 4 parts.
Sulphur precipitate, 1 part.
To one ounce of the ointment add one drachm of potassium sub-carbonate.

Before using the ointment the whole surface of the body, except the face and head, should be well bathed in warm water and castile soap, continuing the bath for at least half an hour, then dry thoroughly with warm towels. After this the sulphur ointment should be well rubbed in, all over the body, saving the head and face.

Woolen underclothing should never be permitted, a patient suffering from *scabies* being allowed to wear only linen or cotton next the skin. After this treatment, the underclothing should not be changed for at least two days, at which time the process should be repeated, with the same preliminaries. All ordinary cases will yield in the course of a very few days, if these directions are carefully followed.

JAUNDICE.

That interference with the regular and natural secretions of the liver involving a general disturbance of the system which is termed "jaundice," is considered by some authorites a distinct disease in itself and by other eminent writers merely a symptom of functional disorders, primarily arising in other organs. Be the theoretic cause what it may, the fact that the liver is not performing its allotted work properly, to an extent that affects almost every organ and tissue in the body, makes the situation a very grave one. There are many opinions as to the reason of this disorder, but they are so varied that an accurate digest of them would be more confusing than otherwise to the lay reader.

Symptoms.

The characteristics of jaundice are infallible and can never be mistaken. The greenish yellow tinge to the skin, whites of the eyes, etc., is always present. The urine and other secretions are dark yellow, often strongly tinted with green or brown. A fully developed case of jaundice, especially if of the malignant type, is a very serious matter and a physician should be consulted before it has made any great progress. Any attempt at home treatment will only end disastrously.

However, it cannot but be admitted that jaundice is frequently no more than a very severe attack of the same complaint whose mild form is known as biliousness. Further, it is always preceded by these less marked premonitory symptoms, and if taken promptly at this stage, the remedies at hand in the household will very probably correct it. Herein is a vivid illustration of the value of immediate action when even apparently unimportant indisposition is evident. A simple bilious attack can be cured in two or three days at comparatively no expense, which, if neglected, may develop into that very serious disease, jaundice, than which there is no harder malady to entirely eradicate from the system, since it penetrates every tissue and even the blood itself. Consequently while we discourage amateur treatment for a pronounced case of jaundice, yet we are certain that many, if not all attacks of this disease, can be prevented by careful attention to the condition ot the system and using every precaution to restore the impaired action of the organs as indicated by the symptoms of biliousness. This statement is of course made with the reservation of those instances where this disease is a concomitant or sequence of fevers, etc., and it is to be applied solely when there are indications of no other ailment than jaundice.

Remedies.

Under its proper head the author has outlined a course of treatment which if followed will correct all bilious disturbances. The remedies prescribed, the *C. D. Co's Liver Pills* and *Sarsaparilla Resolvent* have proven their efficacy in thousands of instances and can be implicitly relied upon.

MEASLES.

Measles is almost exclusively a disease common to children and rarely attacks a person who has reached maturity, although if it has been escaped in tender years, one is always subject to its contagion. While as a rule, one attack insures future immunity from this disease, instances are far from unknown where the same person has suffered from it twice or even three times, at intervals of months or years. There seems to be a peculiar affinity between this disease and hooping cough, epidemics of the latter being followed by measles and *vice versa*.

As measles alone is usually a grave malady and it is very rarely uncomplicated with other disorders, it is imperative that a physician be summoned at once, as soon as the symptoms have determined definitely the identity of the illness. Its highly contagious character dictates that the greatest care be taken to prevent the spread of the disease and all infected clothing, bedding, etc . as well as the apartment itself must be thoroughly disinfected. While the attending physician will indicate the necessary remedial measures which the individual peculiarities of the case require, a few brief suggestions may not be out of order, and will be found below.

Symptoms.

While measles often makes its appearance very abruptly with little or no warning, there are frequent cases where there are marked indications of the advent of some serious disorder, for several days before the rash appears. When communicated through contagion, the period between exposure and development of the disease varies from seven to fourteen days. When the complaint comes on slowly the patient seems to suffer from many of the symptoms of catarrh, sneezing, red, watery eyes, harsh cough, etc., accompanied by fever, which with the feeling of general lassitude increases rapidly until the eruptions appear and the ailment can be definitely recognized.

On the other hand when its advent is rapid, the usual indications are a sudden chill followed by a high fever and within a very brief time by the characteristic rash. Usually after the second or third day of this eruption the fever begins to subside and the patient begins to recover slowly unless complicating diseases set in.

Treatment.

Medicines, except such as tend to allay the severity of the fever, are of little use in simple measles, and little can be done except attend to the wants of the invalid and make him as comfortable as possible. The patient should be confined to the bed and the ight partially excluded, as the eyes are often temporarily weakened in this as in other febrile diseases. Draughts must be avoided as well as exposure of any kind. Frequent bathing as directed in "Hints on Nursing", will prove very grateful, relieving the hot uncomfortable sensations so trying to the sufferer. The diet should be light and nutritious, such as arrow root, milk, chicken or veal broth, beef juice, etc. All the cool water necessary to quench thirst should be given freely, preferably in small quantities at frequent periods, rather than large draughts at one time.

The use of stimulants, as in fact the entire general management of the case, should be left to the judgment of the attending physician.

MORTIFICATION—GANGRENE.

When this distressing condition occurs as an effect of organic disease, which occasionally happens, the previous illness of the patient will have been such that there will be a medical attendant to combat it. Should it result from wounds that have been improperly dressed and cared for, a physician should be called immediately, for the horrible state of death in life is far too serious to be controlled by the housewife.

There is no excuse for the occurrence of gangrene in ordinary wounds in persons whose blood is in a fairly healthy condition. If injuries are properly dressed with a good lotion, such as the *C. D. Co's Antiseptine* they will heal kindly and quickly. Further, those whose blood is known to be tainted with any of the numerous humors akin to scrofula, can renovate and remove the disease from their system if they will. And they should realize that while the impairment may not cause them present trouble, the slightest injury may result in its concentrating itself at that point and nothing bears with it more possibilities of a very serious nature.

We have pointed out many times in this little volume the road by which a perfectly pure and healthy condition of the blood can be reached, and it is the height of wilful folly for any one to knowingly remain diseased, when relief can be obtained at so little expenditure of effort or money.

NEURALGIA.

The popular idea that neuralgia is a local disease is the greatest possible fallacy, as there is no complaint that is more deeply seated throughout the entire system. It is essentially a disease of the nerves and while there exists much diversity of opinion as to its real origin, yet it unquestionably arises from a general debility of the nervous system and anything which tends to impair the nerve vitality must have much to do with the coming of this torturing malady.

The effects of wasting illness, exposure, overwork, insomnia or any of a hundred circumstances that sap the strength and vigor of the nerves is more than liable to bring it on.

It has also been most thoroughly proven that neuralgia with its kindred ailments, is inherited with the imperfect nervous system which they indicate.

Symptoms.

There is not the least possibility of confounding the burning, shooting, tearing pains of neuralgia with any other ache or pain to which the flesh is subject. Its distinctive fleeting character, beginning and ceasing with great suddenness or changing quickly from one part to another, makes its identity certain.

Occasionally there is a sense of soreness to the touch over the seat of long continued neuralgic pain but more often this is absent.

Remedies.

With this as with nearly every other disease, it is impossible to effect a cure unless the treatment begins at the very foundation of the ailment. Local measures may afford temporary relief but the malady will certainly return unless it is thoroughly and entirely removed from its lurking places.

Thus in neuralgia while the application of cloths saturated with *Anodyne Liniment* will almost always ease or altogether relieve the pain, yet immunity from future attacks can be had only by a vigorous course of treatment which will build up the wasted and diseased nervous system.

Rarely indeed is this condition of the nerves unaccompanied by a generally debilitated state of all the tissues of the body and when this is the case, no permanent improvement can be expected until the whole system has been strengthened. This involves the use of a fat or tissue builder, as well as a nerve tonic, and both are a necessity if one would secure rapid and permanent relief.

Hundreds have used those two incomparable preparations, the *C. D. Co's Nervita* and *Emulsion of Cod Liver Oil* in repairing the wasted forces both of nerve matter and tissues, with the most gratifying success. Used together they act on all the impaired parts simultaneously and it is but a question of time when the system will be restored to its original vigor and the shooting agony of neuralgia will be only a disagreeable memory, never more to be feared.

NIGHT SWEATS.

There is no symptom that more surely indicates a state of very low vitality or general debility than profuse perspiration during sleep.

It is an almost invariable accompaniment of the more advanced stages of consumption and similar wasting diseases, and in such instances little can be done beyond the general treatment for the predisposing complaint.

Often however, night sweats are frequently noticed when there exists no acute or marked ailment and as there is nothing more weakening, measures should at once be taken to stop them.

The author has seen scores of remedies recommended for this distressing complaint, many of them being local applications, as lotions, ointments, etc. Nothing can be more absolutely useless than to attempt to correct what is but a symptom of constitutional weakness, by rubbing on some harmless wash.

Night sweats should be taken at their true value, viz;as an indication of general impairment of the vital forces and a momentous warning. Once on the down grade and it will not be long before serious disease will make its presence apparent. Recognize the danger signal, inaugurate a course of energetic and effective treatment to check and replace the wasted vigor and you will be assured of safety.

In cases of this description the *C. D. Co's Sarsaparilla Resolvent* and *Emulsion of Cod Liver Oil* are the best insurance policy that can be had. If you are suffering from night sweats and have a feeling of general lassitude, take the *Sarsaparilla* to tone up and stimulate the organs, and at the same time the

Emulsion to rebuild the impaired tissues, and your return to perfect health will be rapid and certain.

PILES.

Among persons well advanced in years, especially those whose life has been to a great extent sedentary, there are few who have not suffered more or less from this annoying complaint. The predisposing causes of piles are unquestionably luxurious and indolent habits, indulgence in rich foods and wines or other spirituous liquors. The disease is excited to development by any of a hundred circumstances, but most often, by costiveness or other derangements of the bowels.

Symptoms.

The indictions of piles are too generally known to need much comment. The discharge of blood in the form of the disease known as bleeding piles is of course convincing evidence, while the peculiar itching or burning sensation of the little tumors about the anus are indications of the other varieties of the disease where hemorrhage is absent.

Treatment.

Unless the case be very severe and of long standing it can be made to yield quickly to home treatment. Occasionally, through neglect, the ulcers become so large and malignant that only a specialist can remove them, a surgical operation being necessary, but ordinarily nothing is required that cannot be used without the expense of a physician.

The first requisite is to keep the bowels in perfect condition as constipation is the most frequent exciting and aggravating form of the disease. The tendency to costiveness can be easily corrected by taking the *C. D. Co's Liver Pills* in laxative doses, taking care not to get the bowels loose, as that condition is as detrimental as its opposite. One pellet each day will usually regulate the evacuations to the desired condition of having them move easily and freely.

To heal the ulcers use the *C. D. Co's Pile Ointment* applied as per instructions on each packet. It is a guaranteed cure for this disease.

If in addition, the patient carefully regulates his diet and habits, avoiding rich and exciting food and especially wines and liquors, at the same time taking a large amount of active exercise, he can secure relief at little expenses of time or trouble. With immunity so easy, no one need suffer from this most annoying complaint.

INFLUENZA—La GRIPPE.

Notwithstanding the fact there is not a distinct disease that has been more prevalent during the past two or three years in the United States than la grippe, there is no recognized ailment about whose cause and treatment there is a greater lack of definite information. This mystery is unexplainable for while medical history ascribes a rather recent origin to the complaint, at least there are few reliable records of its epidemics until within the present century, yet it would seem as though modern scientific observers should have long since definitely classified it.

Its numerous and varied forms and symptoms have doubtless done much to baffle the medical writer, together with the fact that so many other comparatively unimportant diseases have been erroneously called influenza, particularly when the true type of the disease is epidemic.

The opinion of the best authorities is that true influenza is not necessarily a serious matter when uncomplicated with other diseases. The great danger lies in that it predisposes to other affections, especially pneumonia and weakening the vitality of the system, opens it to the inroads of any other maladies that may be latent. For this reason it should never be neglected.

Symptoms.

Genuine la grippe differs from a common cold being a much graver ailment, requiring very different treatment. The premonitory symptoms differ widely in various cases, but usually the disease is ushered in by sharp fever with an intense headache. It sometimes develops very slowly and again its onset is sudden, in the latter instance being most frequently accompanied by a severe headache, particularly about and between the eyes. Often this fever is interrupted with a pronounced chill, succeeded by the febrile symptoms again.

Either simultaneously or very shortly after the advent of the fever, the catarrhal disturbance begins. This soreness of the membranes of the nose and throat is severe and yields slowly to treatment of any kind, although the cough which accompanies it will usually succumb to a soothing expectorant as *C. D. Co's Cough Balsam.*

One of the marked peculiarities of the disease is the great nervous depression which it induces. This loss of strength and low spirits combined with intense aching of the muscles often seriously effects the patient's mind to extent of severe delirium.

Treatment.

It is always advisable to secure a physician at once rather than attempt the home treatment of influenza or la grippe. While the disease in itself may not result seriously, the liability of grave complications is so great that a skilled practitioner should be in attendance.

Preventive Measures.

While there may exist much doubt as to the cause of influenza, there is not the slightest question that those whose condition is in any way debilitated are much more liable to its attacks than persons of vigorous health. Therefore, it is a wise measure of precaution to have the system in its normal state, when la grippe is epidemic.

If the blood or digestive organs are in the slightest degree out of order restore them with the *Sarsaparilla Resolvent.* Emaciation and general debility are rapidly corrected by the *Emulsion of Cod Liver Oil.* Never forget that this disease is only too often fatal to those who have not the most perfect health with which to combat it. Take time by the forelock and arm yourself against it. The precaution may insure you immunity and will at least lessen the severity of the attack.

PIMPLES—"Black Heads."

These unsightly eruptions or blotches, so common to that period of life which marks the first change from childhood to maturity, are often a source of deep mortification to those afflicted with them. Nor is this condition of the complexion devoid of discomfort not to say disgust for the beholder.

The variety of pimples noticed most frequently, is that known as "flesh worms" or "black heads." It is a very wide spread fallacy that these are caused by minute insects burrowing under the skin and the small bits of round hardened matter tipped with black, which are forced from them by pressure, are often called worms. They are simply the secretions which accumulate and harden in the pores when their natural action is for some cause suspended, the black tips being the dirt which has settled at the exterior end of the duct.

The other common class of eruptions called acne, resemble minute boils, coming to a head and suppurating slightly, then subsiding to reappear immediately in another place unless the predisposing derangement is corrected.

All classes of these facial eruptions are attended with a greasy appearance of the skin and generally repulsive state of the complexion.

Causes.

As the usual time at which these pests are most common is about the age of puberty and a little later, they arise from the general constitutional disturbance characteristic of the great change.

When found in adults they are usually caused by excesses of all kinds, especially in the use of spirituous liquors, indolence, lack of cleanliness, etc.

Remedies.

To remove these unsightly eruptions, the first move must be to reestablish the vigor and tone of the whole system, and purify the blood. All eruptions are infallible evidence that the blood is not in its proper condition and until it is fully restored the pimples will continue.

Take the C. D. Co's Sarsaparilla Resolvent regularly and in a very brief time, not only will a pronounced abatement of the pimples be apparent, but a much improved tone to the whole physical economy can be felt.

Accessory treatment in the way of external applications and exceptional cleanliness are imperative. Baths should be frequent and thorough, and the skin should be briskly rubbed with a rough towel until in a glow. Use nothing but pure castile soap, for one of the frequent exciting causes of disorders of the complexion, is cheap, improperly compounded toilet soaps. The "black heads" should be removed from their lodgement and this can be most easily done by pressing the end of an old fashioned watch key directly over the spot.

To correct the greasy appearance of the skin, bathe the face thoroughly every night in a solution of rain water and borax, about a tablespoonful of the borax to a tea cup of the warm water. Dry with a soft towel then anoint the face and neck thoroughly with the *C. D. Co's Rose Cream.* In the morning again bathe the face with borax water but omit the *Rose Cream* until evening.

If this treatment, including the conscientious use of the *Sarsaparilla Resolvent,* be persevered in, the complexion will in a short time be entirely cleared of the pimples and spots, leaving the skin soft, fair and healthy.

PNEUMONIA.

The very name of pneumonia or inflammation of the lungs sends a greater thrill of dread through everyone than the mention of any other known complaint. Every winter season sees thousands fall before its deadly sweep and other thousands rally from its clutches, only to suffer with shattered health and very early succumb to consumption.

When it is once firmly settled on the lungs, this malady is far too serious to be trusted in any but the most experienced hands and the best available physician should be constantly at hand.

But as a sequence of neglected colds, pneumonia

can very often be prevented if the premonitory indications are treated promptly, as noted under that head. Do not neglect a cold. Nine times out of ten it may run its course without serious results, but the tenth time death may be crowding close upon the wake of the primal symptoms and carelessness will cause life long sorrow,

Symptoms.

The most infallible indication of pneumonia, the hardness of the diseased lung, is not easily determined by an amateur. But the sensation of intense pain through the chest cavity and under the shoulder blade, quick, hurried breathing, fever and restlessness, are momentous symptoms which call for active measures without an instant's delay.

One very marked feature of incipient pneumonia is profuse expectoration of viscid matter of a pronounced greenish or yellow color, often strongly tinged with blood.

Treatment.

As this disease positively demands the assistance of a physician, no directions for treatment will be given, aside from measures which will tend to relieve the patient, pending the doctor's arrival.

As the main reliance of the profession is in strong external counter irritation, no time should be lost in placing a powerful application on the chest. Either use a poultice of the *C. D. Co's Medicinal Mustard* made double the usual strength, or else apply common spirits of turpentine. The turpentine is preferred by many physicians because of the rapidity of its action and it should be used as follows:

Thoroughly saturate a soft flannel cloth with the clear spirits and cover with it the entire affected side of the chest. Over this place two or three thicknesses of flannel wrung as dry as possible from very hot water and if at hand, lay over all a piece of oiled silk. If the oiled silk cannot be had, use several thicknesses of warm dry linen. As the turpentine evaporates rapidly, renew the application at frequent intervals, never permitting the damp cloths to become cool.

While this treatment will prove painful to the patient, causing great irritation to the skin or even blistering, do not desist until the doctor arrives or respiration is slower and easier.

The arrival of the physician will of course terminate the responsibility of the housewife, beyond that care demanded in faithfully carrying out his directions.

For soothing the irritation and soreness of the cuticle of the chest, caused by severe counter irritation, nothing is better than the *C. D. Co's Healing Lotion.*

After pneumonia, as in fact all serious acute illnesses, the patient suffers from great loss of vitality and more or less emaciation. To build up the depleted forces and restore the system to perfect health, always use the *C. D. Co's Emulsion of Cod Liver Oil.* This incomparable renewer of impaired strength should always be kept in every household in the land. It is an agent that one trial will make the constant recourse of the housewife.

RHEUMATISM.

The author deems it unnecessary to devote any of his limited space to the discussion of what rheumatism is. He does not believe there will be any of his readers who have reached maturity but that have sufficient knowledge of its indications to easily recognize this almost universal malady and were he to reproduce one tenth part of the mass of conflicting theory to be found in medical writings regarding its cause, the reader would be inextricably confused.

Suffice to say that rheumatism is distinctively a disease which has its origin in the blood a fact that is admitted by all of the highest authorities.

Its painful and pronounced symptoms are so well known that they will not be referred to in this connection.

Treatment.

Acute cases of either articular or muscular rheumatism call for professional attention if very severe. It is only the milder forms of the acute phase and chronic rheumatism to which the following suggestions as to treatment should be applied.

Obviously the only vulnerable point at which this disease can be attacked with any hope of success, is at its very foundation and starting point—the circulatory system. If it is not eradicated entirely and thoroughly from the blood, it will certainly recur when excited by exposure or any of the circumstances which tend to develop its acute form.

In seeking for an alterative of sufficient power to remove all traces of disease from the blood, the one first turned to is that peerless preparation which has proven its great worth in countless cases, the *C. D. Co's Sarsaparilla Resolvent.* Right here the author wishes to state that this invaluable compound is not claimed to be an infallible *specific* for rheumatism, to effect an absolute cure in each and every instance. It will not do this nor will any product of the chemist's laboratory.

We cannot toostrongly advise its use by those afflicted with this malady, but that there are cases which will not yield to the influence of even this most powerful remedy must be borne in mind, consequently both the author and his prescription should not be condemned, should the trial result in disappointment.

Those who suffer from this disease, which has been so long and universally considered incurable as to have become the very synonym of obstinacy, should willingly try anything that offers even a possibility of cure and should appreciate a compound which only in a degree mitigates their pain.

If all this is considered, we are willing to advise the use of the *Sarsaparilla*, feeling confident that at least ninety per cent of those who follow this suggestion will be more than satisfied with the result. They cannot but receive great constitutional benefit from this unequalled tonic, while more than one-half of the cases in which it is used will be entirely cured of their rheumatism.

As an application to relieve the agonizing pains, the *Anodyne Liniment* will be found very efficacious. It should be rubbed on vigorously two or three times each day and its intensely penetrating action will deaden if not entirely remove the pain.

We earnestly request that those who have tried the scores of so called rheumatic specifics (?) make at least a thorough trial of these standard preparations. While there is a possibility of disappointment, the chances of happy results are very greatly in their favor.

RINGWORM.

This annoying and repulsive disease of the skin is most common among children, but adults are also frequently subject to it.

Its origin is very similar to scabies, being caused by the ravages of a minute parasite and it is nearly always communicated by contagion.

In common with all diseases of this class, the blood is more or less affected and the *Sarsaparilla Resolvent* should be given regularly. The strictest cleanliness should be enforced, bathing the eruptions frequently with borax and water.

Ointment of tar which can be procured at any drug store will be found to have a pronounced healing effect on the sores, or if preferred an ointment of white precipitate of mercury can be used.

SCARLET FEVER.

While this is almost entirely a disease of childhood, adults are not exempt from it providing they have escaped previous attack.

As a rule, those who have suffered from it are insured future immunity, although instances are on record where the same individual has had two or even three attacks of scarlet fever at long intervals. This is a very rare exception however, and one in whom the disease has been well developed and run its course, can come in contact with it with little or no possibility of danger.

It is communicated through contagion only, though certain climatic conditions often seem to favor its development and increase the degree of certainty of its being absorbed by contact, hence the frequent epidemics of the disease, particularly during the autumn months.

Like all the graver febrile affections, this complaint is too serious to be trusted to inexperienced hands and a physician should always be consulted immediately on the symptoms becoming sufficiently distinct to identify the malady.

Therefore the author will merely endeavor to point out the characteristic features of the disease which distinguish it from others of its class and will leave all advice as to treatment and management to the medical attendant in charge.

While scarlet fever is a very dangerous ailment in itself, there are few if any of the children's diseases which are more liable to complication and it should in consequence be watched with the greatest care, and no trouble or expense spared to secure the very best professional aid that can be obtained.

Symptoms.

The period which elapses between exposure and the marked development of scarlet fever varies greatly, ranging from three days to as long as three weeks but rarely more than six or seven days.

The first indication, especially in young children, is a very sore throat with tenderness at the angles of the lower jaw and soreness or stiffness of the neck. This is followed quickly by the symptoms of fever, usually, though not always, beginning with a chill, succeeded by the customary rapid pulse, hot, flushed skin and intense thirst and very often nausea and vomiting.

This stage lasts from twelve to thirty hours when the characteristic rash makes its appearance. This consists of very small dots, bright scarlet in color and often covering the skin so thickly as to leave none of the cuticle of its normal tint. It usually is first noted about the breast, rapidly extending to other portions of the body but it may be first apparent at any point.

By this time or in fact before, the case should have been in charge of a physician and further description is unnecessary.

A final word of caution—the infection of no disease lingers about a room or clothing longer or more obstinately than that of scarlet fever. Every article that has come in contact with the patient must be thoroughly disinfected and on recovery the sick room entirely dismantled, fumigated and renovated.

After a severe attack of scarlet fever, or any serious illness, the patient is always greatly emaciated and in every way weakened. Nothing excels the *C. D. Co's Emulsion of Cod Liver Oil* as a most highly concentrated stimulating food to repair the waste and it should always be given to convalescents.

SHINGLES.

This peculiar zone or belt of eruptions generally about or around the trunk is comparatively rare, but occasionally becomes so prevalent in certain localities as to constitute almost an epidemic.

It is often ushered in by slight feverish symptoms making its appearance as a distinct complaint, but more frequently it occurs as a complication of ague or some severe febrile disease.

It is characterized by irregularly shaped groups or clusters of eruptions, varying in size from half an inch to three or four inches in diameter usually first apparent at about the middle line of the back and extending in a wavy, disconnected band around the right side towards the front of the body. Sometimes another belt or zone will form on the other side, and these cases are popularly supposed to prove uniformly fatal but this opinion is a fallacy.

The course of the disease ordinarily continues eight or ten days, after which if properly treated the eruptions disappear and the patient gradually recovers health.

Treatment.

The first great care in a case of shingles should be not to break the vesicles or small blisters which can be easily noted in the eruptive patches. Their rupture very often leads to the development of obstinate, indolent ulcers very hard indeed to heal. The eruptions should be dressed, either with the *C. D. Co's Healing Lotion* and then thickly powdered with starch, or bathed with a solution of acetate of lead. The former however is to be preferred. After the patches have begun to dry up, frequent bathing with warm water into which a small portion of the *C. D. Co's Antiseptine* has been placed is a most excellent remedial measure.

The bowels should always be kept open and moving freely with the aid of the *Liver Pills*.

The most important measure of all however, is to remove the disease altogether from the circulatory system and doing this most thoroughly and completely is the only way of preventing the recurrence of the malady. Give the *Sarsaparilla Resolvent* regularly from the very outset of the disease, continuing it for several weeks after the acute phase has disappeared and the general health of the patient will not only be better in every way, but the latent poison will have been fully eliminated from the system.

SORE THROAT.

Mild attacks of tonsilitis, pharyngitis or inflammation of any portion of the throat, are nearly always consequences of a cold and if not very severe and accompanied with ulceration, they can usually be controlled by the housewife without professional aid.

But if exceptionally obstinate and when on examination the whitish or yellow patches which denote an ulcerous tendency are very evident, it is always advisable to call a physician, as the extreme form of this complaint sometimes requires a surgical operation to entirely cure it.

For the milder forms however any one of a score of simple remedies will suffice.

The little tablets of chlorate of potash to be had at any drug store will often be sufficient to afford relief. They should be held on the tongue and allowed to dissolve slowly.

A gargle of a strong solution of baking soda and water is sometimes good.

Borax and water is also used.

Another that is highly recommended is salt and water with a good proportion of vinegar.

Steaming the neck or throat, or hot applications are almost certain to relieve the soreness, but great care should be taken to prevent exposure afterward, as the patient is very liable to take more cold.

In all cases of inflammation of the throat, keep the bowels open by using the *C. D. Co's Liver Pills*, for if the outlets of the body are moving freely, inflammation in any part yields far more quickly.

Advice as to curing clergyman's and singer's sore throat will be found under the chapter on "Hoarseness" and if the directions there given are followed, relief and immunity are both assured.

SALT RHEUM.

This very troublesome and inveterate eruptive disease appears most frequently on the hands, but often extends to the arms or even other portions of the body.

It is characterized by patches of minute vesicles or blisters filled with a thin corrosive fluid, accompanied by intense burning and itching. These after a time harden and slough off in small white scales to reappear in a brief period in another place.

This complaint is transmitted by inheritance to a very great degree and it is an exceedingly obstinate malady to entirely cure.

However a persistent use of those two standard preparations, the *C. D. Co's Sarsaparilla Resolvent* and *Rose Cream*, will not only relieve but entirely eliminate salt rheum from the system.

Take full doses of the *Sarsaparilla* regularly three times each day and continue it for a considerable period after the scaly patches have apparently disappeared, if you would drive away the disease forever.

Anoint the eruptions liberally with *Rose Cream*, particularly on retiring and not only will the severe irritation be greatly soothed and quieted, but the powerful healing effect of this delightful emollient will cause the patches to rapidly subside, leaving the skin clear and healthy.

There is a popular superstition to the effect that salt rheum is incurable but this is not a fact. The use of these two compounds will entirely remove it and not only at small expense and little trouble but both are in the highest degree pleasant and palatable. The days of nauseating or disgusting medicinal preparations are past and at the head of modern improvement in this respect will be found the standard products of the *C. D. Co's* laboratory.

SCROFULA.

Because of its wide prevalence, its great obstinacy and the awful straits to which its victims are often reduced, scrofula has become almost the synonym of deep seated, loathsome disease.

In some of its hundred forms it seems to permeate a large portion of the entire body politic and unfortunately, it is apparently increasing rather than diminishing in extent and power.

No taint is more certainly visited upon the innocent offspring of those who suffer from it than this. Scrofulous parents will almost surely beget diseased children and while the malady may be quiescent in one generation, it is very liable to break out in the next, unless the blood is entirely renovated and purified of the hereditary taint.

Consequently all persons in whom there is the slightest suggestion of this awful disease, no matter what their general health or position in life, should never rest until they are *certain* that it has been forever subdued and removed. Not only do they owe this to themselves but it is an imperative obligation to their descendants. What use is there in a life spent in the effort to bequeath their children money and lands, if that priceless gift of health with which to enjoy and use wealth rightly is not given with it. If your posterity can be left but the one, strong bodies are infinitely the more precious than untold millions in bonds or gold.

Symptoms.

The indications of scrofula are legion and the more common forms are known to every one, therefore the author need but refer briefly to them.

The parts which this disease seems to consider its especial prey are the glands, most often those of the neck and throat, although it may affect any or all portions of the glandular system. In these attacks the glands are noticed to become slowly larger and very hard, in the case of the smaller and less important, forming what are often called "kernels" under the skin. This enlargement is usually attended with a considerable degree of soreness although frequently this feature is absent. Unless checked, the glands increase in size, intense inflammation sets in and the swelling finally develops into the most obstinate ulcers known to the medical profession.

Inflammation and ulceration of the tissues surrounding the eyes, is another frequent sign of the scrofulous taint, as are, in fact, very many forms of skin diseases and almost every ulceration of any part of the body.

Wounds that are indolent, refusing to heal quickly and properly, sloughing, suppurating, etc, are certain indications of the presence of this poison in the blood.

In substance, if the blood is in a perfectly pure and healthy state, all injuries will heal kindly and ulcerous disturbances will never be known. If the opposite is noted, then the blood is in a serious condition and alterative measures are absolutely imperative.

That the scrofulous tendency surely predisposes to consumption, is a fact well known to the entire medical profession and conclusively proven by thousands of cases.

The presence of this poison in the system, opens wide the doors to any and every acute ailment, and as it cannot but weaken the vitality of the patient, reduces to a minimum the chances of successful combat with the deadly myriads of disease germs.

With all this considered, it seems impossible that any who know this poison to be latent in their system, should rest contentedly without using every effort and resource to eradicate it. Particularly when this can be done at slight expenditure of either money or trouble. But there are many who are blinded by their temporary immunity and never become aroused until the malignant phase of the disease is fully developed and in consequence recovery is tenfold slower and more difficult, to say nothing of suffering which might have been prevented by timely action.

The author wishes he might warn in tones of thunder, every one who has even a slight suspicion of scrofulous poison in his blood. Strike this demon while it is sleeping. Slay it in its lurking place and thus insure to yourself and yours a future of health and happiness.

Remedies.

The noted specialist, from whom after much solicitation and an immense outlay in money the *C. D. Co.* finally secured that invaluable formula the *Sarsaparilla Resolvent,* was widely known for his phenomenal success in treating scrofula in all its forms and phases. And this invaluable compound is the weapon with which he laid low this most persistent devil of disease.

While unexcelled as a general alterative, having a quick and most pronounced beneficial effect on all ailments which originate from humors in the blood, this wonderful preparation was originally designed and used as a specific for scrofulous affections.

Tens of thousands of successful cures testify to its value as a specific for this malady and there is not a well developed case of this dreadful affliction that will not yield to its potent action. Take it regularly three times each day in full doses and it will require but a brief period to demonstrate the truth of this statement of its virtues.

Do not think however that even this most powerful remedy will cure in one day or one week. There is no malady which more thoroughly and completely involves every organ and tissue of the system and naturally it will not relinquish its grip without resistance. The fight must be constant and unflinching to subdue it. It may take months to entirely eradicate the last trace of scrofula from the blood, but the certain reward of faith and perseverance is sure to come and the patient be freed from its oppression for all time.

Patients in whom scrofula is thoroughly developed, invariably suffer from emaciation and wasted vitality. Such should always take the *C. D. Co's Emulsion of Cod Liver Oil* in connection with the *Resolvent*, for unless the system has sufficient strength to aid the *Sarsaparilla* in its action, the poison will still linger, despite the most powerful remedial measures.

Ulcers and scrofulous sores should be dressed with the *C. D. Co's Antiseptine* and *Medicated Absorbent Cotton*. They not only have a strong healing effect but their antiseptic properties are of infinite value in preventing malignant blood poisoning.

TYPHOID FEVER.

If one-tenth part of the degree of apprehension

with which all classes regard cholera were given to typhoid fever, the result would be a great diminution in the vast amount of those who every year fall victims to its ravages.

It is but another example of the truth of the saying that "familiarity breeds contempt," for while there may exist a universal dread of this disease, it is not held at its true value because of everyone having become more or less accustomed to its presence. Typhoid is one other of that class of diseases which have their origin in, and are transmitted by, germs or bacteria. The common impression that typhoid or enteric fever is contagious, has no foundation in fact, but that it is often epidemic to a greater or less extent is caused by the presence of vast numbers of these minute microbes, which are always absorbed through the mouth and stomach into the intestines, the seat of the malady.

At certain periods these bacteria swarm in waste places, especially poisoning sources of water supply and it is in this manner, more often than any other, that they secure lodgment in the human system. They breed and multiply in the same homes as all their death-dealing fellows, viz; cess-pools, closets and all receptacles for refuse matter. Exterminate and keep them down with the constant use of disinfectants and your household is guaranteed safety from this dreaded fever. When typhoid is abroad in the community, use the *C. D. Co's Germ Killer* every day in all out-houses and every possible lurking place of germs, and immunity for you and yours will be the reward.

The best method of curing disease, is to prevent it and more than one-half of the sickness of the human family could be avoided by the intelligent application of those means which scientific research and commercial enterprise have placed at the disposal of every one.

While typhoid is not of itself necessarily fatal, yet the peculiar liability to serious complications which attends it, stamps this disease as one of the most grave and difficult that the profession is called upon to treat, and places it entirely beyond the management of the housewife. When the symptoms of indisposition are at all indicative of typhoid fever, a physician should immediately be summoned.

Symptoms.

The advent of typhoid is very insidious, the patient declining in health so gradually that it is often impossible to fix the exact date at which pronounced illness was first noticeable. For days previous to the severe stage of the ailment, there is lassitude, loss of appetite and more or less headache.

This phase is followed by the first indications of the acute disease, nausea and vomiting, great restlessness, diarrhœa, furred tongue and usually high fever, etc, etc. Bleeding at the nose is common and the bronchial cough very rarely absent.

If he is not already in attendance, the calling of a physician should not be delayed and the responsibility of the household healer will, of course, be transferred to him. The individual peculiarities and complications of a case of typhoid fever are so varied, that no reliable directions for treatment or diet can be given. This must be left to the knowledge and experience of the physician.

One point—all excreta from a typhoid patient is charged with the poisonous germs of the disease and should therefore be treated with the *Germ Killer* before being disposed of. Pour two or three tea spoonfuls of the clear liquid into the receptacle, then empty in a remote place.

. After a siege of this wasting fever, the patient is invariably greatly reduced and to restore flesh and strength give him the *Emulsion of Cod Liver Oil*. It will build up the waste tissues with marvelous rapidity and hasten immeasurably complete recovery.

TOOTHACHE.

Often the intense pain through the teeth, jaw or face is of a neuralgic origin, aggravated either by exposure or a decayed condition of one or more of the teeth. But usually a common toothache is caused by decay or ulceration at the root of the tooth and can only be permanently relieved by a dentist.

Temporary alleviation of the pain can, however, be had by the use of one of the following remedies: The *C. D. Co's Toothache Drops* are a specific in almost every instance and should be used in preference to anything else, when at hand. If these cannot be had, try a few drops of oil of cloves, chloroform or oil of sassafras. Any of these should be applied by saturating a small bit of the *Medicated Absorbent Cotton* and filling the cavity with it.

If of a neuralgic character, the pain can very often be quieted by wrapping a hot iron or brick in a cloth saturated with equal parts of *Anodyne Liniment* and water, and thoroughly steaming the face, neck and head.

Toothache is an indication of some disturbance or derangement of the structure of the teeth and a good dentist should always be consulted at the very first opportunity, no matter if the pain is temporarily subdued.

TUMORS.

These swellings or enlargements are almost infallible evidence of scrofula and should be heeded accordingly.

The surgeon's knife will and should remove them, but unless vigorous alterative treatment corrects the diseased condition of the blood, they will surely return and possibly at a point that will entail grave results, such as about the ovaries, etc.

The author has endeavored to give great emphasis to the importance of removing the scrofulous taint from the blood, in the chapter devoted to that disease, and would refer the reader to that portion of this little work.

Never attempt to remove even small tumors unaided by professional assistance and advice. All operations are exclusively the province of a skilled surgeon, and no amateur should ever risk even apparently unimportant matters of this kind.

ULCERS.

These serious suppurative sores may arise from any of a score of causes, but, as has already been stated, they almost invariably come from a diseased state of the blood and where the blood is perfectly healthy they are rarely, if ever, to be found. The exciting cause may be a wound or bruise, acute scrofula, fever, syphilis, etc.

To cure them it is absolutely imperative that the blood be purified or renovated by a course of *Sarsaparilla Resolvent* and until this is done no dressing or external application will effect more than temporary relief.

In combination with the internal treatment as above, use the *C. D. Co's Antiseptine* and *Medicated Absorbent Cotton* in dressing the ulcers.

Saturate sufficient of the *Absorbent Cotton* with the *Antiseptine* to completely fill the orifice and cover the immediate region of the sore to prevent the exuding pus from spreading contamination to the surrounding parts, and firmly bandage with soft linen. Renew every twenty-four hours, washing off the hardened matter from the surface about the edges, with a solution of warm water and the *Antiseptine*, but not bathing the interior of the cavity.

This dressing will not only keep the suppuration free and healthy but will in a very short time entirely heal the most obstinate ulcers, providing the *Sarsaparilla* is taken regularly as directed.

WARTS.

These unsightly eruptions occurring chiefly on the

hands and most common in children, are of no consequence aside from their being a disfigurement.

They very frequently disappear of their own accord, especially with the growth and development of a child, but if it is considered desirable to remove them, it can be done by any of the following methods:

Burn the top of the wart every day or every second day with lunar caustic. Apply in the same manner, with a small brush, either carbolic, sulphuric or acetic acid.

Another very effectual plan of removing them is to tie a silk thread very tightly around the base of the wart as close to the skin as possible. This ligature will gradually kill the wart, causing it to slough off.

Occasionally these pests are very persistent, reappearing again and again in the same place as fast as removed, but usually one thorough treatment as above will eradicate them permanently.

WORMS.

The presence of worms or parasites in the intestines is a very frequent cause of indisposition and general disturbance of the entire system. This is especially true of childhood and there are few mothers who do not have to wage a more or less constant warfare with these interlopers, during the growing years of their family. Consequently the symptoms of their presence are well known to the majority of women and the author will dwell but briefly on them. The ordinary long, round, and thread worms are referred to in this connection, tape worms being a graver matter and one requiring professional aid if the case be at all severe. Unimportant instances of tape worm however, are practically included under this head, the symptoms of the milder cases being similar and the remedies prescribed effectual, except in patients where the tape worm is very large and firmly seated.

Symptoms.

Probably the only certain means of definitely determining the presence of worms in the intestinal canals, is by noting either the parasites themselves or the ova which pass from the patient in the process of defection.

The usual constitutional symptoms, are defective nutrition, in other words the patient appears to derive no benefit from food, often pain in the abdomen, nausea and pallor. Other indications may be swollen eyelids, irritation of the nostrils, grinding the teeth during sleep, etc.

Any of these may denote the presence of worms and as the remedy is pleasant and harmless, measures should at once be taken to remove the parasites from the system.

Remedy.

Give the *C. D. Co's* standard *Vermifuge* in doses proportioned to the age of the patient, as directed on the bottle.

This is one of the safest and at the same time most

thoroughly efficacious preparations known to the *materia medica* and it should be constantly in every household where there are children. It has strong soothing and tonic qualities and its use will prove most highly beneficial in every way.

MEMORANDA.

Every housewife is constantly accumulating formulas, recipes and bits of valuable information of various kinds, which it is in the highest degree important to preserve. Memory cannot always be relied upon implicitly and it is a wise measure to always write these out in full and thus guard against all possibility of error or loss through forgetfulness.

Small memorandum books, almanacs, etc., are very liable to become mislaid and to supply this need for a permanent place for this data, where it can be always at hand and in a convenient form for ready reference, the following pages have been left blank and ruled especially for this purpose.

In the table of contents at the beginning of the book will be found blank lines to correspond with these pages and if the private recipes are indexed, they can be found without an instant's delay.

The publishers have substituted this plan in the place of the great array of miscellaneous rules, formulas, etc., which are usually to be found in a work of this character, few of which are ever used, believing that the arrangement which permits the preservation of one's own recipes, which have been proven valuable, will be more generally acceptable.

www.ingramcontent.com/pod-product-compliance
Lightning Source LLC
Chambersburg PA
CBHW030354170426
43202CB00010B/1368